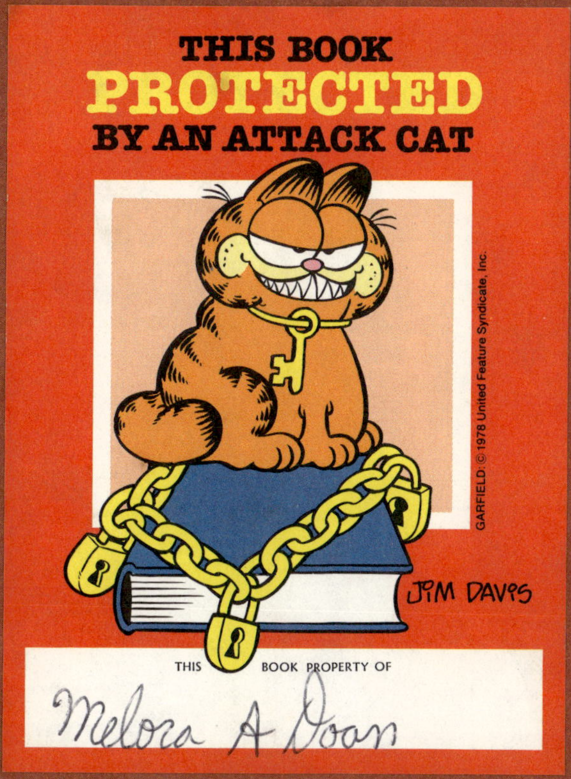
THIS BOOK
PROTECTED
BY AN ATTACK CAT
GARFIELD: © 1978 United Feature Syndicate, Inc.
JIM DAVIS
THIS BOOK PROPERTY OF
Melora A Doan

A *Changing* World for Wildlife

by George S. Fichter
illustrated by Jo Polseno

GOLDEN PRESS • NEW YORK
Western Publishing Company, Inc.
Racine, Wisconsin

Produced in the U.S.A. by Western Publishing Company, Inc.
Published by Golden Press, New York, N.Y.
Library of Congress Catalog Card Number: 76-25330

Contents

BRONTOSAURUS

Introduction

SUDDENLY the world is alarmed about animals becoming extinct. Why? Extinction is not new. It is a natural process that has occurred as long as there has been life on earth. Scientists estimate that for every species alive today a hundred others once existed and have since become extinct.

Over the milleniums, living conditions on earth have changed drastically many times. Each time a great change took place, animals not suited for survival in the new environment died out and were replaced by species better equipped for the altered conditions. But as we reckon time, such changes came about slowly—over millions of years. Natural extinctions are thus measured on the grand scale of geologic time rather than in the short span of man's recorded history.

Dinosaurs provide a classic example of an entire group of animals that disappeared by natural extinction. For 120 million years dinosaurs were the dominant animals on earth. There were hundreds of species; some were no larger than chickens, others were the most fearsome beasts that ever trod the land. The largest dinosaur measured 90 feet from snout to tail and weighed 100,000 pounds. Another stood 20 feet tall and walked on its hind legs. But 65 million years ago, the dinosaurs disappeared. With them went the fishlike ichthyosaurs, plesiosaurs, mosasaurs, and other reptiles that ruled the seas.

What happened? During the heyday of the dinosaurs, the earth's climate was warm even in the polar regions. Palm trees grew as far north as Greenland. Then the earth slowly began to cool. Great ice caps formed at the poles and pushed toward the equator. The earth's crust warped into craggy mountain ranges that replaced swampy lowlands. For the giant reptiles, fitted for life in a warmer, different landscape, the earth became intolerably hostile. One by one, the reptiles became extinct and were succeeded by warm-blooded animals better suited for life in the changed world.

This kind of extinction has been repeated many times in earth's history as conditions for life have changed. About 500,000 years ago—which is recent in geologic history—most of the earth basked in a warm climate.

Then, over a period of thousands of years, a great change took place. The snows were deeper, the winters longer. In the polar regions, the snow and ice became heaped in places to as much as two miles deep, and slowly these big masses began creeping southward. At times the glaciers covered most of the Northern Hemisphere. Over about 400,000 years, the great glaciers advanced southward four distinct times. The final and longest period of glaciation lasted 70,000 years and ended only about 10,000 years ago.

With each period of glaciation, the major life forms changed. Some kinds simply retreated southward in front of the glaciers and did not return. Others disappeared, and new kinds that were better adapted to the colder climate took their place. Notable among the animals that became extinct were the woolly mammoth, woolly rhinoceros, and cave bear, but many smaller, less spectacular animals also failed to survive these drastic natural changes in the world climate.

Large animals have greater difficulty adjusting to changes in the environment than do smaller animals. Similarly, animals with generalized needs are less affected than are those that have specialized needs. Also, animals that produce numerous young are more likely to bridge the changes than are those that produce few young. The greater the number of offspring, the greater the chance for the occurrence of individuals able to adapt to the new conditions. Thus the unspecialized and prolific cockroaches of today are much like those that lived during the Coal Swamp days some 400 million years ago. Scientists point out that cockroaches are likely candidates for survival even if the earth becomes virtually unlivable for other forms of animal life.

But if extinction is a natural process, why the current concern? One reason is the general awareness now that extinction does occur. But more important are the proliferation of new factors that cause extinctions, the accelerated rates at which exterminations take place, and the possible dire consequences to modern man.

Worldwide, the list of animals threatened with extinction includes nearly 1,000 species. About 200 of these animals inhabit North America. Since the discovery of the continent, 62 animals have become extinct—57 of them in the last hundred years. Since 1900, annihilations have continued at the rate of one animal per year. These are all large vertebrates. Smaller animals are not at present documented, but they are becoming extinct, too, as man continues to manipulate and to poison the world. Conservationists estimate that on a worldwide basis only about 1 percent of the endangered animals has been listed. This is principally because the inventories of smaller and "lower" animals are difficult to make. One list was swollen recently by the addition of almost 200 species of butterflies. Even without adding these animals, however, the lists are almost of frightening proportions. In the United States, for example, roughly 10 percent of all the vertebrates or animals with backbones are now either rare or endangered.

Some extinctions go unnoticed and may be of such little consequence that the loss does not matter. A few years ago, for example, a new species of a small moth was discovered in an area south of Miami, Florida. Collectors began making regular visits to get specimens, but suddenly there were none—the species was gone. What happened to it? A change had taken place in the area. To get rid of weeds and to hasten the natural succession process toward the establishment of a basic stand of grass in the grove lands, owners had been mowing regularly for several years.

The most noxious of the weeds were the sprawling, gangly white-flowered plants that produce sticktights or Spanish needles. This plant happened also to be the specific food of the moth. Mowing had destroyed the moth's "little world." Few in number from the start and existing in a very limited range, this particular moth was short-lived as a species. Its loss probably makes no difference to the total natural world. The numerous similar species that live in the area quickly filled the vacancy. Similar kinds of extinctions probably occur regularly as nature experiments with slightly varied types.

New species of animals arise as a result of changes in inheritance factors. These changes often begin with offspring referred to as "sports" or mutants. Most mutants do not survive, for usually they are not as well equipped for survival in their environment as are their parents, organisms already tested by time. But of the many mutations that do occur, an occasional one is an improvement. If the change produces a transmissible characteristic, a new species results. Even a subtle change may be enough to alter the direction taken in adapting to new living conditions and may in time produce a new species.

In the sudden and total extinctions caused by man, all of the hereditary materials harbored by a species developed over millions of years are irreplaceably lost. Particularly with the larger birds and mammals, these losses are unfortunate because the animals might have contributed significantly to man's welfare.

As one example, the great auk was literally clubbed into extinction because it was so good to eat. If it had been preserved as a domesticated or even a semidomesticated bird, it could have fed many people, especially the increased number who might live in polar regions in the future. The great auk got its food from the cold arctic seas. Our domestic fowl cannot survive in that region. Was there space for the auk to live? Certainly. Greenland just recently created the world's largest national park—an area of 270,000 square miles, virtually uninhabited by man but offering nesting sites in abundance for the great auk.

Several species of wild cattle now nearly extinct might either be developed into domestic breeds themselves or be used to produce hybrids with qualities surpassing those of any we now have. Wild stock should be spared whenever possible, for it may be necessary to turn again to these hardy base gene pools if the domestic breeds developed from them over the years do not satisfy needs or are unable to survive as conditions change. We cannot anticipate future circumstances, but we can prepare for them by preserving original stocks.

The importance of a particular species of wildlife to man's well-being may not always be direct and obvious. In Africa, the governments of several countries recently authorized the killing of hippopotamuses that were destroying crops. Each hippo was eating more than a hundred pounds of crops every day. But when most of the hippos were exterminated, a strange and unexpected thing happened: the numerous tilapia, a fish that lives in streams and lakes, became scarce, too. The tilapia prospered on the thick growths of algae that grew in waters fertilized by the excrement of the hippos. Man in turn suffered because the tilapia were his principal source of protein. No one had suspected that there was a relationship between the hippopotamus and food for man.

Numerous similar instances might be cited. In many cases, for example, man has systematically eliminated pests of crops and then discovered that at the same time he has destroyed the natural controls over the pests. The population of the pests rebuilds quickly, with no checks over growth. This frequently happens, too, when animals are introduced to a new environment. Out of their natural surroundings and without control elements, the animals commonly "boom" in numbers and become pests rather than welcome additions, often threatening native species by competing with them for food and living space.

We share the world with all living things, and as the dominant and thinking creatures on earth, we have become their custodians. We have no special right to deprive any form of life of continued existence. But above all other reasons, including pure esthetics in some cases, we must be concerned about extinctions —particularly the kind occurring now—because they represent changes in the natural world of which we are a part. We are responsible for the differences in world environments that have made them unlivable for many animals. Each animal that becomes first rare, then endangered, and finally disappears from earth shared basic needs with us. Extinction brought about by upsets in natural balances or that resulted from impoverishment or poisoning of the land, air, and waters are blatant warning signals. We can be an endangered species, too, unless we learn to live in harmony with nature.

None of the original major habitats, or biomes, is the same as before man began to explore and to populate the world. Some have been totally altered, and all have been changed at least to some degree to suit man's needs. The chapters that follow describe the earth's principal life zones, some of the changes that have taken place in them, and how man's intrusions have affected the original animal populations. Most of the reports of man's effect on the environment are repetitious accounts of the dead and the dying and of man's savagery as a predator. Often they reveal him as having little more wisdom than the grasshopper in Aesop's fable.

But fortunately, this is an accounting of what has happened in the past—from which we learn the lessons that guide us in the future. A reparation has been started, and the list of animals "saved from extinction" may help erase the shameful document of those now endangered.

GREAT AUK

SEAS OF GRASS

BISON—North American prairie

VAST seas of wild grasses once rippled in the wind at the core of every continent. They were cropped regularly by the great herds of grazing animals that inhabited these lands separating the dry deserts from the forest lands. Companions to the grazing animals were the flesh eaters that preyed on them, the scavengers that finished off the remains, and the hordes of smaller animals filling every niche of the grassland world. In temperate regions, the original grasslands are gone, converted mainly into fields of grain that produce food for man. Wild grasses still grow in the open lands of the tropics and the subtropics, but there, too, the cinch draws tighter for wildlife as man finds ways to put these lands to use for his own needs.

The Great Plains

Once the heartland of North America was a great prairie—the Great Plains. From the evergreen forests of Canada to the deserts of Mexico and from the Rockies eastward almost to the Mississippi, a green, waving sea of grass grew on the gently rolling land. On the eastern side, closest to the Mississippi and its tributaries, the grass grew tall—to as much as ten feet. It graded to tough, scrawny, shorter grasses toward the dry eastern slopes of the Rockies. This was before the white man usurped the Indian.

Here, at some times of the year, the Plains Indian could put his ear to the ground and hear the thundering hoofs of the stampeding herds of buffalo, or bison. These sprawling grasslands were the home of an estimated 60 million of the shaggy beasts, the great bulls weighing more than a ton and measuring six feet tall at their shoulders. In autumn, the herds moved southward to warmer territory and a winter food supply. In spring, they traveled northward again as the grass turned green with tender growth under the warm sun.

Buffalo were the Indians' mainstay in this treeless, largely waterless land. The buffalo ate the grass, and the Indians ate the buffalo. The hides of the buffalo were turned into various garments, or they were stretched over poles to make teepees. Tools, weapons, and ornaments were made of the bones, horns, and teeth. The dried, pancake-shaped dung became the buffalo "chips," burned as a fuel for heating and cooking in this expansive, treeless land where wood could be found only along the few watercourses. No part of the killed buffalo was wasted. Tons of meat were dried to provide food for the tribe until the next herd passed. Often the meat was mixed with wild seeds and berries to make it last longer.

On foot, the Indians could not keep pace with the speedy buffalo galloping off at 20 or 30 miles an hour. The Indian hunters had to kill quickly and then either track down the herd again or wait for its return. Sometimes they set fire to the dry prairies to force the buffalo to pass the waiting hunters. When the Spaniards brought horses to the New World, the Indians then could follow the buffalo. They killed more than in earlier days. But even with this added advantage, the Indians did not dent the buffalo population.

It was the white hunter who whittled the herds into near oblivion. With their rifles and then with trains to haul away the meat and hides to eastern markets, professional killers began to reduce the herds at an alarming rate. After the Civil War, the hunters carried breech-loading rifles, replacing the old muzzle-loaders. With these fast-action guns, a hunter could fire many times before a stampeding herd was out of range. An experienced hunter could kill as many as a hundred animals in a day.

Unlike the Indian, the white man was wasteful. He took only the best parts of the slaughtered animals. The great hump was an especially prized portion, as was the thick, dark tongue. But the prices paid for even these select pieces of buffalo meat were ridiculously low—usually less than five cents a pound. Tongues sold for twenty-five cents each. Robes made from buffalo hides cost as little as a dollar.

The prairie became littered with the rotting carcasses of slain animals, which were sometimes shot from train windows simply for the "sport" of killing and with no intention of utilizing the animals. By the early 1880's, after less than half a century of commercial hunting and with most of it in the years following the Civil War, the great herds of buffalo were gone. Those in the southern half of the prairie country went first. Then, as the railroads cut through the prairie land to the north, the buffalo vanished there, too.

The last big year for the buffalo hunters was 1882, when nearly a quarter of a million hides were shipped to market. Shipments then dropped suddenly to fewer than 500. No other large animal has ever been so cruelly, systematically, and speedily reduced in numbers to near annihilation. The killing was allowed and even encouraged by government officials because elimination of the buffalo was the surest way of driving the Indians from the region and opening the land for white settlers.

Reduced in number from 60 million to about a thousand animals, the buffalo was on the verge of extinction. Some 600 of those last remaining animals were herded onto a federal reserve. There, fortunately, the animals bred and prospered. Small herds are kept today on government-managed lands, their populations held in check by controlled harvests. The buffalo was thus spared the fate of becoming only a name in a book.

Meanwhile, the white man continued his invasion of the grasslands. Cattle and sheep grazed where the buffalo had once roamed. Fires burned off the tall grass, and plows bit deeply into the dark soil to turn it for planting fields of grain. There was no place here for the buffalo as the sea of grass became the breadbasket for a growing nation of people.

Roaming the grasslands with the buffalo was the pronghorn or American antelope, its total population before the coming of the white man estimated to be nearly 40 million. To escape an attacking wolf or coyote, this fleet grazing animal could dash off at 40 miles per hour or even faster, according to some observers. After this burst of speed to outdistance its pursuer, the pronghorn would slow down to 30 miles per hour for several miles before stopping to resume its feeding.

But this remarkable speed was no security against the gun, the fence, and the plow. Along with the buffalo, the pronghorn was pushed into more remote

PRONGHORN

BLACK-FOOTED FERRET

regions. Though it is much reduced in numbers—probably to about 250,000—the pronghorn does persist today. Only a less common subspecies is endangered: the Sonoran pronghorn that lives along the border of Arizona and Mexico. Smaller and paler than those found farther north, only about three dozen of these animals exist today. They were either killed off by hunters or were unable to compete successfully with domestic livestock on the ranges. An added complication has been the spread of the animals' range over the Mexican-American borders, calling for international controls to protect the few remaining animals of this distinctive subspecies.

Gone from the grasslands now is the plains wolf, a subspecies of the timber wolf. The wolves preyed on the young or on the sick and dying buffalo and pronghorns, or they ate with as much gusto the smaller animals. But with this abundant fare eliminated, the wolves sometimes turned to calves, sheep, and other domestic animals for their meals. Their archcompetitor and enemy then became man; and with guns, poisons, and fences, man won the battle.

The most abundant of all the four-footed animals in the grasslands was the black-tailed prairie dog. Its total population was beyond counting, but several

BLACK-TAILED PRAIRIE DOG

hundred million of these social burrowers probably inhabited the grassy prairies. Prairie dog "towns," each consisting of a thousand or more animals, dotted the grasslands; one huge "city" was estimated to extend over 30,000 square miles.

Prairie dogs policed themselves to maintain a respectable balance in numbers based on the space available. Each family marked off its needs in the prairie, and when an errant animal from another group wandered off limits and entered foreign territory, he was promptly whistled at until he scurried back to his own family. But prairie dogs rarely ventured more than a hundred feet or so from their burrows. A greater distance from the entrance to a burrow was too dangerous, for out in the jungle of grass were coyotes, wolves, and other predators. When the buffalo came, they trampled down the grass. Then for a time, the prairie dogs were safer because they could see farther. When danger came their way, they gave off shrill "alarm" whistles that sent all of the animals scurrying into their burrows out of sight.

The worst foes of the black-tailed prairie dogs were those that went directly into their burrows, which were intricate mazes of underground passages that helped to keep the soil of the plains well plowed and aerated. Burrowing owls lived in abandoned holes or in little-used passageways. They took their share of the wandering young and the older, less alert animals. Black-footed ferrets preyed on the prairie dogs regularly, as did the badgers. Prairie dogs were victims, too, of the prairie falcon. The little rodents never learned to look up for this danger that swooped in on near-silent wings. But the prairie dogs were so abundant that there was a plentiful supply of food for all their natural enemies.

The white man became the black-tailed prairie dog's most dangerous enemy. The prairie dog's burrows were a menace to horses, which could easily break a leg by stepping into one of the holes. The mounds and the holes were not appreciated either by the plowmen, nor were the animals themselves when they stuffed their stomachs with field and vegetable crops. Driven away by the ranching and farming activities, the prairie dogs disappeared like the buffalo. Only a longer time was involved in their disappearance, and the less spectacular size of the animals detracted from the magnitude of the happening.

Black-tailed prairie dogs are found today only in limited areas of the southern sections of the plains where the killing of the "varmints" still continues. These once-plentiful animals are now rare. Disappearing with the prairie dogs are those animals that depended on them for their survival: the black-footed ferret, the prairie falcon, and the western burrowing owl. Badgers have survived by moving into other regions. Their dependence on the prairie dog was less direct.

Other animals also border on extinction now that the great grasslands of North America are gone. Among them are Attwater's greater prairie chicken and also the lesser prairie chicken. When the prairies were fenced and plowed, these henlike birds were forced into the spotty remnants of their former habitat. There the already reduced numbers were subjected to heavy hunting. The birds were often shot purely for sport, with the dead left lying where they fell. Today the populations of both species are at dangerously low levels. In the early days, a million or more Attwater's greater prairie chickens lived in the southern region of the grasslands. Now there may be fewer than a thousand. Saving these native grouse can be accomplished only by providing protection for them on extensive ranges of grasslands. There the cocks can boom and strut

ATTWATER'S GREATER PRAIRIE CHICKEN

WHOOPING CRANE

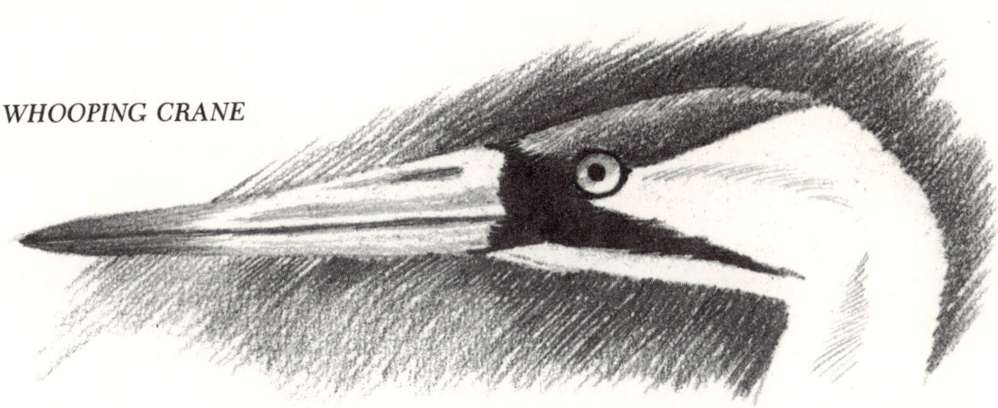

for their mates in early spring, and the hens can shepherd their chicks in reasonable safety. Such lands should soon be set aside.

Gone forever from the prairies now is the giant whooping crane. The last of the birds—about 50, the number fluctuating only slightly over the past 25 or 30 years—spend their summers in Canada and winters along the gulf coast of Texas. The teetering population could easily be eliminated by a single disaster. That possible fate came to the fore in the spring of 1975 when nine of the remaining whoopers migrating northward to their breeding grounds made a landing on Funk Lagoon on the Sacramento Game Management Area near Kearny, Nebraska. Federal and state biologists had expected the birds to land there, a traditional stopping place, and had assembled scaring devices and brought airplanes in for use in scaring the birds off. They wanted to force the whooping cranes to continue their flight to some other area. But the birds came in on a day of heavy, freezing rain. The planes were grounded, and no one saw them arrive.

Why the concern in 1975? Cholera, one of the most dreaded and deadliest diseases of birds, was rampant in the area. The disease had already killed more than 15,000 waterfowl. Further, it appeared to be worst at Funk Lagoon where several dozen ducks were dying every day. This one fateful landing seemed to destine 20 percent of the whooping crane population to sure death. It would take years to build the population to the same level again, and there would always be a chance of some similar disaster. But having tipped the balance to this precarious low, man is obliged to do what he can to bring the big bird back. Conservationists remain determined.

The birds could not be scared off, but after a day and a half of rest, they resumed their journey voluntarily. They were apparently unaffected by the disease, which biologists consider quite miraculous. Those infected ordinarily die within 24 hours.

No one knows how many smaller kinds of animals that were peculiarly adapted to the prairies also became extinct as a result of losing their abodes and a supply of their kind of food. Perhaps none of these extinctions would make life different for man, who has turned the great sea of wild grasses into a sea of cultivated grain to feed himself and his domestic animals. But all are testimony to man's powerful effect on changing the world of wildlife.

The African Savannas

No land areas on earth have hosted more big animals than the African savannas —two broad grassland belts, each about 600 miles wide, on either side of the equatorial rain forest. They are bordered both to the north and to the south by deserts. Near the rain forests, the grasses grow tall, with a few scrubby trees scattered through the flat land in places. Approaching the drier regions, the grasses are short, and the land is treeless or has only a sprinkling of thorny shrubs. In all, the savannas originally occupied about 4 million square miles or nearly 40 percent of the great African continent.

This is still the home of more than 40 species of large mammals. Here can be seen single herds of as many as 10,000 wildebeests and zebras. Elands, impalas, kobs, topis, bushbucks, gazelles, waterbucks, kudus, sunis, giraffes, rhinoceroses, water buffalo, warthogs, elephants, lions, cheetahs—all are beasts distinctly identified with Africa.

For many centuries, literally millions of these animals lived in the savannas in balance with nature. The people of the land hunted some of them for food and for hides, but the hunters were too few in number to affect the total population. Problems started with the arrival of the Europeans, who began to colonize the savanna areas about 300 years ago. Ever since, living conditions for wildlife have grown worse year by year.

As the land was settled, roads began to ribbon the wild areas, and more and more of the open lands were fenced for livestock. Bit by bit, the living space available for wildlife was reduced. At first the remaining animals could retreat to still-open areas, but this is no longer possible. Once a traveler might have seen as many as a million animals of different species in a day. Now he may go for hundreds of miles without seeing even one. In many areas he is likely to see more people than wildlife.

Wise conservationists recognized the potential threat to Africa's wildlife almost half a century ago and were fortunately successful in getting a few large

LIONESS AND CUB

WHITE RHINOCEROS

tracts of land set aside as sanctuaries. But man is pathetically plodding when it comes to correcting his own errors. More positive steps must be taken soon to save some of the many kinds of imperiled animals of these vast grasslands. For some, whatever action is taken will be too late to save them from extinction.

In Garamba National Park, primarily a grassland, large numbers of elephants and square-lipped or white rhinoceroses were making a healthy comeback until a war in the early 1960's brought several years of slaughter that again reduced their populations to critical lows. In other areas, similar local wars have reduced populations of wildlife.

The largest and most spectacular of the parks in the savannas is Serengeti National Park in Tanzania. Covering about 25,000 square miles, it is the home of the largest remaining herds of grazing animals. Protected largely in government lands, too, are the now rare cheetahs, the most doglike of all the cats and also the fastest of all land animals in short-distance runs—to 60 miles per hour or perhaps even faster. These cats do well in captivity but do not produce young readily. For this reason, it is all the more important that those in the wild be protected. Unfortunately these magnificent cats have also been hunted by man. Their spotted pelts are much prized, and the hunters sense some sort of accomplishment for having killed one of the beasts. So hunting combined with destruction of their habitat and also their food has brought the cheetah to the brink of extinction.

Among the other animals rapidly becoming rare are several kinds of antelopes. Among these is the great sable antelope, noted for its exceptionally large horns—to as much as five feet long—as well as being hunted for its flesh. The great sable antelope once ranged over most of western and central Africa, but today it is confined to two areas of Angola. Only a few hundred animals survive. Both areas where the great sable antelope lives are restricted to hunting, and the populations of the animals have been climbing steadily. Poaching does occur regularly, however. The giant sable antelope is not yet safe from extinction.

In addition to the shrinkage of living space, a substantial share of the devastation of Africa's wildlife has been due to hunting—not, however, to get food and hence justifying at least to some degree the slaughter. The prime objective for most of the killing junkets has been to acquire trophies—the head of a lion, a rhino's horn, an elephant's tusks, the foot of an elephant to be fashioned into a wastebasket. The killing for curios has indeed been shameful. The amount involved is not known, since most of these animals are protected by law. The killing, selling, and exporting continues, however. From Kenya alone, as one example, more than 200 tons of ivory from elephant tusks finds its way to markets annually. Many animals killed for sport only have been left where they fell, to become food for hyenas and other scavengers. Senseless slaughter of this sort satisfies only the perverted hunters.

Inventories of wildlife in the African savannas, as elsewhere in the world, are made mostly of animals in their present ranges. It is commonly ignored that many of these animals are already existing in greatly shrunken ranges that do not offer optimum conditions and that their numbers are much reduced. Often forgotten, too, are the animals that have already disappeared from the savannas.

In the African savanna, for example, there once existed a zebra called the quagga, so named because its nasal-like neighing sounded to the native people like a drawn-out or wheezed *quag-ga.* Sometimes the sound was described as being more like the baying of a dog. Unlike other zebras, the fleet quagga had stripes only on the front half of its body; the remainder was yellowish brown. Though never among the most common of the savanna animals, the quagga did roam the southern grasslands in herds of thousands.

European settlers hunted the little horselike quagga both for its flesh and its hide, from which they made shoes and also bags for carrying grain. As in other similar examples, the killings were more in the nature of massacres than hunts, and by 1800 the quagga was beginning to be scarce. The last of these animals in the wild fell before a gun about 1879. A female in the Amsterdam Zoo lived until 1883, then died without leaving offspring as had all the others ever kept in captivity.

The white-tailed gnu or black wildebeest may soon become the quagga's companion. Once common from the Cape to the Transvaal, the white-tailed gnu stands less than four feet tall at the shoulders. With its bristly mane and tufts of hair on its nose and chin, it gives the appearance of being a ferocious animal—which it can be, unusual for members of this group. More than a century ago it was hunted almost to extermination for its meat and hides. Now protected on reserves, in zoos, and on some private lands, the herds appear to be building again, from a low of only a few hundred animals to possibly several thousand. But the population is not yet secure.

LION

Nearly all of the animals of the African savannas are much reduced in numbers, in fact. This includes the lion—the "king of the beasts." The plight of the lion was dramatically evidenced a few years ago when a Hollywood movie producer making a film of Africa felt obliged to import several lions from wild animal farms in the United States to get beasts that looked convincingly regal.

Poaching on land reserved for wildlife has become another major threat. No one would be inclined to prosecute poachers if they killed to get meat to supplement a meager diet. But in nearly all cases, the poachers make their kills for absurd reasons. A black rhino's horn, for example, commands a very high price. It is pulverized and sold as a medicine believed by some people to have unusual or mystical qualities as an aphrodisiac. Elephant tusks also bring a high price for making carvings and other useless curios. Wildebeests are sometimes killed for only their bushy-tipped tails to be used to shoo away flies. Thousands of animals are killed every year for such trivial purposes. Rather than using noisy guns, the poachers commonly catch their quarry in snares or in pitfalls.

Still there are some places in the African savannas that can boast of as many as a hundred big animals per square mile. In addition, there are many kinds of smaller creatures—literally hundreds of species of rodents as well as skunks, meerkats, civets, servals and other small cats, snakes, tortoises, and countless insects. These smaller animals, as in the other major habitats on earth, demand less space, hence are in less immediate conflict with man's designs on the land. Most of them are adapted to survival where the winters are dry, the soil becoming so crumbly that it is picked up in whirling clouds of dust by the slightest breeze. With the coming of summer rains, the land greens again.

When the land is dry, most of the larger animals migrate to river lowlands or go into the forests, then return when the grasses grow. Several species of baboons, vervets, patas, geladas, and some other monkeys go into the savannas but make their homes in the forests. Giraffes feed in the areas where trees grow, foraging in the high branches that are well above the reach of the shortnecked gazelles, impalas, zebras, wildebeests, and other grazers.

Herbivores must be on the alert at all times for the stealthy, swift attack of a cheetah, a pride of lions, or some other predator. Elephants, protected from predation by their giant size, are destructive feeders, uprooting many trees and shrubs as they forage. They take only a small portion of the vegetation themselves, but other animals then move in to feed on what the elephants have brought down to their level. Elephants must have large tracts of land on which to forage, and with so little now left for food and cover, the remaining herds of elephants have become small. The females do not bear young at as early an age nor as often as they did in years gone by. In some areas, elephants are dying of malnutrition. On government lands, the herds of elephants and other animals

AFRICAN ELEPHANT

are systematically reduced to the carrying capacity of the land, but how long the elephants can maintain themselves in this much smaller world is indeed a question.

Many kinds of birds also inhabit the savannas. These include several species of weaverbirds that are noted for their intricately built nests, and the giant ostriches—to eight feet tall, the largest of all living birds. Ostriches, now threatened with extinction, can run 40 miles an hour, outdistancing pursuers. But, if cornered, they can be dangerous foes, slashing with the two clawed toes on their powerful legs. Adding full complement to the bird population are the carrion eaters—the vultures and storks—as well as such aggressive hunters as tawny eagles and secretary birds.

All of these animals are specially equipped to survive the usual hazards of nature. The changes of the seasons, the predators—these are part of their natural world to which they have adapted gradually over the countless centuries of their existence as a species. Man's intervention has been an abrupt catastrophe. It is at least somewhat comforting to know that man now recognizes how great his impact is on nature. He is making efforts to correct mistakes and to alter his ways. Still there are people who are dominated by avarice and who are blindly unconcerned about what happens to world environments. Until they learn, the existence of many of the mammals, birds, and other creatures that make the African savannas so spectacular will continue to hang in the balance.

The Plains of Southeast Asia

A triangular peninsula jutting into the ocean and separated from the Asian mainland by the lofty Himalayas, India is one of the most heavily populated of all nations in the world. Only remnants of forests remain over vast areas that have been converted to growing crops or grazing animals. Over much of the area, rains seldom come, and short grasses grow on the dry lands not regularly plowed and planted, most of them worn out from thousands of years of use. Many animals that once lived in the forests as well as invaders from the adjacent deserts have become adapted over the years to this man-created grassland.

Survival anywhere in this land of almost 600 million people is difficult, especially for larger animals. It is most remarkable that so many have managed to do so. The cheetah once roamed these plains. It was here that it gained its reputation as the hunting leopard. Young animals were often captured and tamed, then taken into the plains with hoods over their heads. When a herd of blackbucks was sighted, the hood was slipped off, and the swift, anxious cheetah moved off to make its kill, usually rewarded with a drink of the slain animal's blood. Then the hood was put over the hunter's head, and it was hauled away in a cart to its enclosure. As long ago as the 1930's, the cheetah was listed as extinct in Asia. A few animals have been seen since then, but there appears to be no space left for these big hunters.

Rapidly disappearing, too, is the cheetah's favorite prey, the fleet blackbuck. Male blackbucks have twisted or spiraled horns nearly 2½ feet long, and their face is attractively marked with white, brown, and black. Known for their high, stiff-legged leaps when startled, blackbucks can outrun the cheetahs in long distances, but they cannot outrun bullets. And like the cheetah, they can no longer find living space. The blackbuck unfortunately is becoming extremely rare over most of its range. The nilgai and the four-horned antelope are faring better only because they retire to wooded areas for protection. Rarest of the large animals of the plains is the kouprey, a wild ox, found only in the grassy forests of Cambodia. Most authorities estimate its total population at about 150. Koupreys stand more than six feet tall at the shoulders and have horns about three feet long frayed at the tips in bulls. As in other regions, the animals suffering least in the Asian grasslands are the smaller kinds that do not compete as directly or as severely with man for food and living space.

CHEETAH

MANED WOLF

The South American Pampas

In South America, the core of grasslands is primarily the pampas, a treeless plain that, in Argentina, extends from the Atlantic Ocean on the east to the foothills of the Andes. Where the grasslands move into tropical Brazil, they are called campos; in Colombia and Venezuela, llanos. Most of these grasslands are used now for growing grains or for grazing cattle. Only the most hostile areas have been left to wildlife. These are principally lands where the temperature is too high, the rainfall too little, and the winds too strong to suit man's needs. Rodents and other small animals escape the ravages of nature by spending most of their lives in burrows. Birds and larger mammals migrate when conditions become too uncomfortable.

Rheas are tall flightless birds, the South American counterparts of the ostriches of Africa and the emus of Australia. They can travel long distances with ease and commonly graze with cattle on the open ranges, thus inhabiting the pampas all year. Only recently has their conflict with man and his personal interest in the land been great enough to put the birds in danger. The small, grouselike tinamous also appear to be safe at the moment, but their population must be monitored regularly. Herons, ducks, geese, parrots, and many other kinds of birds appear in the pampas during the warm wet months but leave when the grasslands become cold and dry.

More than 20 species of armadillos live in the South American grasslands. The nine-banded armadillo is the most wide ranging, occurring over most of South America and northward through Central America to southern United States. The most typical inhabitants of the pampas are the fairy armadillos, less than six inches long. Becoming extremely rare, these little armadillos have only a partial shell, perched on the back and head almost as though it were added as an afterthought. Fairy armadillos spend most of their life in a burrow, using it not only as a cool retreat but also as protection from predators. The much larger hairy armadillos, so called because of the hairy bristles that beard their armor, are the most common of the armadillos on the pampas.

Armadillos are primarily insect eaters, though they may sometimes eat other small animals, if they are abundant and easily caught. Wherever armadillos become numerous, they are usually persecuted—treated as pests even though they do no actual damage beyond their habit of digging burrows that can be bothersome in some places.

The pampas fox, probably the slowest moving of all the foxes, and the now endangered maned wolf, usually a solitary hunter rather than hunting in packs as other wolves do, are two of the common predators of the pampas. The maned wolf has a foxlike head and a distinct mane, but most outstanding are its long stiltlike legs that lift it high in the grass for a better view in hunting. Like other predators neither species is abundant, hence they can easily become extinct.

BURROWING OWL

The pampas deer is already so rare that conservationists are alarmed about the survival of the species. As the pampas grass on which it fed and in which it took refuge disappeared, it became easier for man and other predators to find the deer's hiding places. Several areas have been set aside as sanctuaries, but the population may have passed the point of recovery.

Burrowing owls inhabit the burrows of viscachas, tuco-tucos, and other rodents of the open country. Wherever these animals or the communities of which they are a part come into conflict with man's needs, the wildlife suffers. Now, while there is still a richness of wildlife in South America, is the time to heed the concerned conservationists. Death knells need not be sounded in most instances. Unlike the governments of Africa, the nations of South America have been slow in setting aside parks of their native grasslands as wildlife reserves, just as they have established only a small number of parks elsewhere on the continent.

The Grasslands of Australia

About 40 percent of Australia is semiarid, offering only scrubby vegetation and grass as food for its animal inhabitants. Still more of the interior of this smallest and next to the last of the continents to be discovered is a desert—accounting for another 40 percent of the land area. Areas of ample rainfall, with climates ranging from temperate to subtropical, fringe the "down under" continent's coasts on the east, southeast, and southwest.

Living in Australia is the most unusual assemblage of animals in the world, the result of the continent's 60 million years of isolation that divorced these creatures from the mainstream of evolution. Most mammals of Australia, for example, belong to the primitive groups—the monotremes and the marsupials.

Monotremes are the only mammals that lay eggs. They are represented by the Australian platypus or duckbill, which lives in freshwater streams and lakes, and by two species of echidnas or spiny anteaters. Echidnas are not as rare as the platypus but still are not abundant. They are in need of protection to prevent more decline in their population. Echidnas are burrowers, making their home mainly in forests but commonly moving into the grasslands in search of ants and termites. The common Australian echidna has spines almost as formidable as those of porcupines. This, fortunately for the echidna, has also made its skin useless to man.

Marsupials, the other group of primitive mammals represented so abundantly in Australia, carry their young in a pouch or marsupium on their belly. The

ECHIDNA

young are born prematurely, their development completed while they are in the pouch. In the absence of competition from the more advanced placental mammals, the marsupials have prospered in Australia, adapting to every environmental niche. There are kinds that complement in appearance and habits the placental mammals that are dominant in other parts of the world. Thus, the Tasmanian devil fills the place of the wolves in the Northern Hemisphere; kangaroos occupy the same habitats as the large grazers in the grasslands of other continents; flying phalangers take the place of flying squirrels, wombats are like the marmots and other large rodents, and numerous kinds of smaller marsupials are similar to rats and mice. Other marsupials burrow underground like moles.

The most advanced of the mammals, the placentals, were also represented in Australia when the continent was discovered by Europeans. There were a number of species of bats and rats, as well as the dingo or wild dog that was presumably carried to the continent centuries ago by visitors from nearby Pacific islands. Many other animals have been introduced to Australia in the years since the continent's discovery, and some of these have come into conflict with native animals, competing for food and living space.

Partly due to the settlement of the continent but also much as a result of hunting and collecting (to satisfy curiosity), Australia's unusual animals have suffered greatly. Some of them, including various kinds of kangaroos, now verge on extinction. Members of the kangaroo family range in size from some that are not much larger than big rats to the giant gray and red kangaroos that may stand nearly six feet tall. In all there are about 50 species in the family. Over a dozen of these are endangered.

Rat kangaroos, sometimes called tungos in their native land, are much like jerboas, with hind feet longer and larger than the head. In most areas they are now rare or are disappearing rapidly. They seem to be holding out with the greatest success on some of the less populated islands. Once there were at least five species on the continent; now only one survives, living in companionship with the introduced rabbits. The partnership is of no special advantage to the rat kangaroos, however. The aggressive rabbits take their share of the food first, and the rat kangaroos get what is left. Neither gets an abundance, for most of the food goes to the cattle and other grazing animals. Additionally, the rat kangaroos have been greatly reduced in number by the poisons spread by the settlers who are concerned with eliminating "vermin," the pest category into which they put almost all small animals.

As recently as the 1800's, hare wallabies were common in the grasslands of southern Australia. They got their name from their remarkable resemblance in size, appearance, and habits to hares, and they have the amazing ability to jump to a height of eight feet or more. Unfortunately, these unique little animals are now extremely rare, as are several other wallabies. They were unable to compete with the domestic animals on the grazing lands, and many were also exterminated by hunters. Among the most successful are the several species of rock wallabies that have fared better because, while they may go into the grasslands to graze, they have retreated from the open lands as a place to live. The rocky hill country has become their stronghold.

The giants of the kangaroo clan are the red kangaroo, a sleek and beautiful animal that lives mainly in the open country, and the eastern and western gray kangaroos, which are found most abundantly in or near forested regions. The largest of these animals measures about eight feet from the tip of its nose to the end of its thick, muscular tail. It is a fact—and not said unkindly—that kangaroos are considerably less intelligent than the more advanced placental mammals. They are fleet, but they have the unfortunate habit of not keeping on the move, making an innocent and fatal mistake of stopping to look back when pursued. At that instant they become an easy mark for a hunter.

Kangaroos are so distinctly identified with Australia that it would be presumed all efforts might be bent toward preserving these animals. To the contrary, kangaroos have been mercilessly slaughtered to get their hides for making purses, wallets, shoes, and other leather items. They are also killed because ranchers complain about their "plague" abundance, though kangaroos are so rare in most areas now that their damage is not great. Kangaroos can generally subsist on the coarser grasses that domestic livestock will not eat.

Year after year a million or more kangaroos are slaughtered, taking the animals further along the route to extinction. Protective laws may have rescued some species, but for others, it may already be too late. Bans on the export of kangaroo hides or the products made from them would help. Other countries should rally to the aid of the animals by refusing items that may get out of the country through black markets. It is true, of course, that in herd numbers both the kangaroos and the smaller wallabies can be damaging to crops. They stuff their stomachs with food that could fatten livestock. Fencing discourages their invasion, and because of their much-reduced numbers, the inroads made into the larders of man and his domestic animals rarely pose a serious threat. Controlled harvests to keep the population of kangaroos in balance will probably be the ultimate solution, but this can come about only when the kangaroos are provided with a suitable and spacious sanctuary.

How long kangaroos can survive in their threatened world is questionable. Because conservationists have rallied to their protection, the larger species appear to be safe at the moment. Watchfulness cannot be relaxed, however, for there are always political groups that influence government policy and that do not see the value of the kangaroos.

Another of the unusual Australian marsupials is the stump-tailed, almost bearlike wombat—a heavy-bodied animal with a large head and short, stubby legs. Wombats have powerful claws for digging. They live in burrows and come out at night to forage on plants. The hairy-nosed wombat that lives in the grasslands is now an endangered species over most of its range. It has been killed off as a pest and also hunted for its pelt. Like the kangaroos, the wombats could easily make a comeback if given adequate, enforced protection.

EMU

As on other continents, the birds of Australia's grasslands live a precarious existence wherever man has designs on the same land for use either in grazing sheep and cattle or in raising grain. Paralleling the ostriches of the African savannas and nearly as large are the flightless emus. Once there were three species; now there is only one, which is greatly reduced in numbers. Standing to almost six feet tall and weighing more than a hundred pounds, the emu was hunted to get meat for the table, and the eggs were also collected and eaten. Despite the extensive hunting, which continues even today, and even after an all-out "war" against the birds by the government in efforts to exterminate them, emus still exist. If emus are to survive, now is the time to establish sanctuaries for them. Even though the emu is the national bird of Australia, efforts to exterminate them still persist. In 1974, for example, resolutions were introduced to the government by sheep and cattle ranchers to annihilate these birds.

A number of kinds of smaller birds, many of them never abundant, now teeter on the brink of extinction in Australia. Among these are two species of scrub-birds: the rufous and the noisy. Both live in the scrub at the edge of the grasslands; both are victims of reduced living space as a result of settlement. Fewer than 50 rufous scrub-birds are believed to exist today. Also rare and endangered are several species of parrots adapted to life in the grasslands.

The Steppes of Eurasia

Too far inland to be moderated by the oceans, the steppes of Eurasia are seared with heat in summer and are bitter cold in winter. For centuries longer than the Great Plains of North America, the Eurasian steppes have been mostly under cultivation, constituting the great grain belt providing food for man. Extending from the Ural River on the east to the Ukraine on the west—roughly 2,500 miles wide—the steppes vary from sparsely wooded areas in the north to open grasslands and prairies on the west and semideserts in the south. Even in densely populated Asia, few areas of the steppes are heavily populated.

Over most of the steppes the rainfall is not more than 30 inches annually, and in much of the thousands of square miles in the southern region, it is much less. Scattered thickets of trees and shrubs grow in the wetter areas, but over the region generally, there is little vegetation to offer protection from the harsh elements. Food is scarce, and water even more so. The few wild animals remaining in the steppes congregate where water collects, and as these places dry up, they move on in search of others. For some the only source of water is the plants they eat.

The most successful mammals of the steppes are the rodents, such as hamsters and gerbils in the very dry areas and the susliks, jerboas, and marmots in other

areas. By burrowing beneath the surface, the rodents escape both the intense heat of summer and the extreme cold of winter. When food is abundant, they store surplus supplies in their underground chambers. Many of the rodents hibernate in winter, and they may also become dormant during the summer. There are, of course, predators that feed on the rodents. These include corsac foxes, wolves, steppe polecats, tawny eagles, and various hawks. The larger animals migrate from the region when the temperatures become intolerable.

As in other parts of the world, it is the larger animals that have suffered most in the steppes because of their more direct competition with man for food and living space. Lack of cover and meager food supply have made coexistence with man virtually impossible for most species.

One of the animals that may have disappeared in the wild in recent years is Przewalski's horse. It is possible that a few dozen still roam freely, but some authorities believe the horse, which once roamed the steppes in sizable herds, is now extinct. It has been preserved in captivity, however, with more than a hundred in zoos through the world. These are the offspring of about two dozen animals that were captured in the early 1900's after the elusive wild horse had been rediscovered by Count Przewalski in 1881. The disappearance of the animals in the wild is attributed partly to hunting but is due mostly to the animals being unable to compete with domestic livestock for food and, more importantly, for water in the dry land.

PRZEWALSKI'S HORSE

Przewalski's horse, which has an erect, brushlike mane, is believed to resemble closely the kind of primitive horse that thousands of years ago moved across the bridge of land connecting North America with Asia. From a dog-sized browsing animal to a large grazer on the grasslands, the horse evolved on the North American plains, but it also disappeared there. There were no horses then in North America until they were reintroduced by the European explorers.

Other large animals that lived on the steppes in the past have been gone for so many centuries that there is no longer a trace of them except as recent fossils. Until about 1600, the steppes had a large population of aurochs, which were large wild cattle. These were exterminated by man, and the European bison, almost as large, nearly followed them into extinction. As has happened with several other animals, the bison owe their preservation to zoo specimens that bred in captivity. The offspring of these animals were put on reservations, and there are now more than a thousand of these bison, their herds controlled so that the population does not exceed the carrying capacity. In historic times, too, lions roamed these vast plains. Presumably it was the same species of lion that is found in Africa today, but with it, surely, was a remnant population of the cave lion, which coinhabited Eurasia with man in the years just following the glacial periods.

Saigas, a kind of antelope, persist in small numbers on the steppes, the herds moving southward in winter and then returning to the grazing lands of the northern regions in summer. Except for man, their only natural enemies are the wolves, which are now also rare. In the 1930's, it was believed that the saiga was about to become extinct. Now, aided by protective laws (though some commercial hunting still goes on), the saiga's total population is near a million.

Birds have suffered, too, in the steppes. Less than 40 years ago, a subspecies of the African ostrich lived here, but it is now extinct. Great bustards, standing to 3½ feet tall, were once common here but are now rare. Somewhat turkeylike in general appearance, the great bustards can run rapidly but fly clumsily. Again it is these larger animals that are affected first and most obviously by man's intrusion into their world.

The natural worlds of Eurasia were the earliest to be destroyed, their animals exterminated or driven into remote regions. But happily, restorations are being made in these same regions, with numerous preserves and all of the native animals that still exist.

WATERLESS WORLDS

KIT FOX—
North American desert

Deserts are dry lands, the annual rainfall less than ten inches. In hot deserts, the temperature commonly rises to more than 100 degrees during the day and may soar to 150 degrees in heated pockets. Nights are cool. In the cool deserts of mid-latitudes, the temperature is cool during the day and becomes bitter cold at night. Based on the availability of water to living things, much of the tundra and also the polar regions are deserts.

Despite these seemingly unlivable conditions, a number of kinds of animals thrive in deserts. And when the rains do come—commonly in downpours—deserts quickly burst forth with an amazing variety of colorful flowers. Plants must produce their fruit and seeds in that brief period when water is available to them. During the remainder of the year, they may be dormant, or nearly so.

The Sahara, largest of the world's hot deserts, extends across northern Africa from the Atlantic to the Red Sea. It is an area equal in size to the United States. On the eastern side of the Red Sea, the desert continues through Asia Minor as the Syrian and Great Arabian deserts and is connected through northern India to the cool Gobi Desert of Mongolia. Only a portion of this massive desert is sand, but in these areas, the sand may be piled to depths of several hundred feet. Swept by the wind, the sand lies in crescent-shaped dunes, like giant waves in the sea. In other places, the strong winds have scoured the earth clean, leaving only gravel and bare rock.

Two other deserts occur in southwestern Africa, the Nambi and the Kalahari, which is noted for its foglike mists. A large portion of southwestern United States and the west coast of Mexico are desert regions, too, and South American deserts include the Atacama along the Pacific coast of Chile where the rainfall is less than half an inch per year, the Gran Chaco in western Argentina, and the caatinga in eastern Brazil. Most of central and western Australia consists of a desert, uninhabitable by man except along its fringes.

Probably the best known of all desert animals, because they have served man over so many centuries, are camels—the two-humped Bactrian camels of the

Gobi Desert and the one-humped Arabian camels of the Sahara. A few Bactrian camels are believed to exist yet in the wild, but they are rare. All of the one-humped or Arabian camels are domestic animals. Even those few that are found in the wild are feral beasts.

Like other desert animals, camels are specially equipped for their life in the dry habitat. Despite the common belief and the wealth of folklore supporting the tale, camels do not store water in their hump. Rather, it is a storage place for fat, principally a food reserve though it may supply some water as it is broken down for use in the body. The camel has little or no fat elsewhere in its body. In other animals, a layer of fat helps to insulate the body and hold in heat, but in camels, it is important to radiate heat from the body. Long eyelids protect the camel's eyes, and its nostrils can be closed tightly with valves—both features that help to keep out sand. The camel has only two toes on each foot, and the toes are connected by skin, thus providing a flat surface that prevents the animals from sinking into the sand.

For many centuries, the camel was man's only dependable means of transportation in the desert. The camel caravans moved slowly. Trips that can be taken now in only a few days with motor vehicles required several months in those days. But it was the remarkable ability of the desert-adapted camels to go for long periods with little or no water that made the journeys at all possible. On long trips, a camel might lose nearly a fourth of its body weight, principally in fluids, but when water became available, it would drink 20 gallons or more without stopping, quickly regaining its weight.

Large animals are rare in deserts, but some gazelles and antelopes are adapted for life in the dry regions. They rarely drink water. Addax antelopes get water from the leaves and roots of the plants they eat. Other large animals stay near waterholes, moving off to find a new one if the one they have been using dries up.

Among the gazelles that are well adapted for desert life are the Moroccan dorcas and Cuvier's gazelle. Both have become rare. Only small herds exist now where once they were abundant. They are victims mainly of modern weapons that have been put into the hands of native peoples.

Most desert animals are small and cannot travel great distances. These include the many kinds of reptiles and also such invertebrates as spiders and insects. They utilize water sparingly, obtaining all of their needs from succulent plants or from the body fluids of the animals they eat. Their wastes are almost totally solid, the water being absorbed from the wastes for reuse in the body.

Many of the smaller desert animals live in burrows in which they escape the intense heat during the day. They become active at night when the desert cools. Birds, because they can fly high above the heated desert floor are the most active animals during the day. They take refuge in the hollows of cacti or in rocky crevices. Most desert mammals—and particularly those that do not burrow—have much larger ears than do their nearest relatives in temperate regions, just as those that live in the Arctic have comparatively very small ears.

The big ears are well supplied with blood vessels, acting as radiators to allow body heat to escape. Also, of course, they help to detect both prey and predators in the open spaces. Desert mammals typically lack sweat glands, for it is important to them to preserve body fluids. Nearly all desert animals are pale, colored much like the sand or rocks on which they live. This not only camouflages them but also reflects heat rather than absorbing it.

In recent years man has moved deeper and deeper into the deserts, irrigating the arid lands and turning them into highly productive agricultural lands—at least temporarily. To accomplish this, water is pumped from deep wells, lowering the water table and thus reducing the already critical supply. More water is pumped out than is replaced, and so the growth of native plants, the basic food and in many cases also the water supply for desert animals, is eliminated. Eventually, in most cases, the land no longer produces crops and is abandoned. Left behind are residues of salts and other chemicals that continue to keep the desert from returning to its original condition.

Even where crops have not been planted, large tracts of arid to semiarid lands have been fenced for grazing livestock. This again has destroyed the basic habitat for both animals and plants. The giant cacti, symbols of the American Southwest, are hosts to a community of desert animals that can only continue to exist if the saguaros are preserved. The saguaros grow very slowly. Often plants as much as ten years old are still only a few inches tall. In natural conditions, they are protected during these early years by growing in a cover of native shrubs or grasses. But on grazing lands, the saguaros are destroyed by the cattle or sheep. As a result, the tiny saguaros are eaten along with other cacti by rodents, or they are trampled by the grazing animals.

KANGAROO MOUSE

Everywhere man has invaded he has killed without thought of preservation. One of the world's endangered species is the Arabian oryx, a beautiful antelope that has been hunted to near extinction. The most devastating killing has taken place in recent years. With machine guns fired from moving trucks, whole herds have been exterminated on single killing sprees. Hunters with rifles have been no less destructive. They have also used motor vehicles to keep pace with the fleet, darting animals, gunning them down one at a time—a feat actually requiring not much more skill than shooting ducks in a rain barrel. By the early 1960's, the Arabian oryx appeared to be doomed. Some estimates placed their total population at about 200 animals; others at fewer than a hundred.

Zoos in London, Phoenix, and Rome as well as in Saudi Arabia began building herds from the few animals that could be trapped. An international conservation organization launched a special program to provide animals for the zoo in Phoenix, Arizona, where the climate resembles most closely what the animals are accustomed to in their native land. Fortunately, the Arabian oryx thrives well in captivity. From fewer than two dozen animals in zoos in the 1960's, the number has climbed now to nearly 150. It appears that the Arabian oryx is assured of survival—in captivity if not in the wild.

The African wild ass also became a victim of modern guns. Native herdsmen accused the wild ass of eating grass that they wanted for their goats, and so they set out to destroy these little animals, the progenitors of the domestic

ARABIAN ORYX

donkey. They did not take into consideration that the domestic goat is one of the most destructive of all animals in the world, doing much more damage to the environment than the native wild ass.

In most cases the wild asses were simply killed to get rid of them. Less often they were slaughtered also for food. The pity is that these little animals are not appreciated as a natural resource, able to thrive where other animals cannot—until man intervenes. Both races of the African wild ass—the Nubian and the Somali—are reduced in numbers now to a few hundred, and in some areas their identity has been lost as a result of interbreeding with the domestic donkey.

In the American desert, even a fish has been threatened with extinction by man. The pupfish, a living relic of a group once common when these lands were flooded with seas, lives in the warm, saline waters of Devil's Hole in Death Valley. When the few hundred fish there seemed imperiled, President Harry S. Truman added Devil's Hole to the Death Valley National Monument to give them protection. In 1973, however, a rancher owning land adjacent to Devil's Hole began a heavy irrigation program, flooding his land from deep wells. As he pumped, the water level in Devil's Hole dropped steadily. Finally the government ordered him to stop irrigating until the water returned to a level safe for the survival of the fish.

Gila monsters and their nearby relatives, the Mexican beaded lizards, are the only poisonous lizards in the world. People could not resist killing these sluggish, slow-moving lizards, rationalizing their act by saying they had eliminated a dangerous animal. But because they are becoming scarce, the lizards are now protected by law in Arizona. Similarly, several desert-dwelling lizards, snakes, and tortoises are now either rare or endangered as a result of shrinkage and destruction of their habitat. The tortoises in particular have suffered because many are caught and sold as pets, usually not living long in captivity where conditions do not match those of their natural environment.

Hunting, trapping, and poisoning have put the big-eared kit fox of the American Southwest close to extinction. The kit fox learned to escape the desert heat by burrowing into the sand, sometimes to depths of five feet or more, but it did not learn that man is more sinister than the sun. Bighorn sheep were once abundant and widespread in the subdeserts and mountains of western United States and Canada, their total population estimated at close to a million animals. Their numbers have dwindled to only a few thousand. Bighorns are surefooted in the rocky, mountainous terrain, making them safe from coyotes and other natural predators, but their agility did not get them out of the reach of man's guns or make them immune to the diseases spread by domestic livestock. In addition, much of their original habitat was destroyed. The bighorns were near annihilation, but they are now making a comeback, protected on a special preserve.

In all of the world's deserts, animals live perilous lives because of man, who is never content to let the living continue without molestation.

ONCE A WOODS

WHITETAIL DEER—
North American deciduous forest

THE VIRGIN forests of deciduous trees, once widespread in the temperate regions of North America and Eurasia, long ago fell to the saw and the ax. With them went many species of wildlife, for this has become the most heavily populated region of the world. Here are most of the cities and towns, and the open areas between are either farmed or broken up into smaller living tracts. Space for most of the kinds of wildlife that inhabited the region originally no longer exists. But the change has not been bad for all animals.

In North America, for example, whitetail deer are more plentiful now than before the land was settled. Whitetail deer do best in cutover land. Brush and young trees put browse food within their reach. When the land was first settled, deer were really not abundant in the virgin timber areas. Hunting reduced their population to a critical low—probably to as few as half a million animals by about 1900. Then the forests began to grow back into brushlands, and as this happened, coupled with protective laws, the deer began to increase in number, too. Despite the many more hunters now, the total population of whitetail deer is probably about 10 million. In some areas, deer have become so abundant that controlled hunts are organized from time to time to reduce the population to what the land can reasonably support.

In China, an unusual series of events saved a deer from extinction. About 1860, a French missionary, Père David, learned that some animals considered to be sacred were kept in a royal game preserve near Peking. The preserve was surrounded by a high stone wall. When he asked for permission to enter the preserve and was refused, the missionary bribed the guards to allow him to climb the wall.

Inside was a herd of more than a hundred strange deer. He was told that these deer could no longer be found anywhere in the wild, and so he was determined to get specimens to send to scientists in Europe. Once again he bribed the guards and soon had two skins to send to Europe. Then the Chinese government also relented and sent three live animals, which unfortunately died during their journey. The skins and the specimens were enough to establish Père David's deer as a new species.

Zoos throughout Europe soon began to clamor for specimens, and the Chinese government responded with more than a dozen deer sent to various zoos for exhibiting. The deer bred well in captivity, and so the zoos soon began selling the offspring, keeping only a breeding pair. Meanwhile, disaster struck the herds in China. It came first as a flood that swept away the wall of the ancient preserve and allowed the herd to escape. Only a few dozen could be rounded up after the flood subsided. Not long afterward, all of these animals were killed and eaten by soldiers during a rebellion. One lone female survived until 1920.

By 1920, too, all of the deer kept in the zoos in Europe were dead. Like many other animals, they seemed unable to survive as only pairs. Fortunately there were still Père David's deer in Europe, however, because one wealthy sportsman had been buying the "extra" animals over the years to establish a herd on his private estate. He began returning his surplus animals to the zoos to give them a new start.

Today there are hundreds of Père David's deer in zoos and preserves, but these strange deer with their long shaggy tail, goatlike hoofs, and large antlers like those of reindeer would not be alive today except for the insistence of the missionary that he be shown what sort of animals were kept beyond the walls of the royal game preserve. No one knows how these deer lived in the wild originally, but it was undoubtedly the encroachment of man as the land was cleared that eliminated them from their natural habitat.

Some animals, such as the red fox, have done surprisingly well in very heavily populated areas, even in cities. Wolves, bears, wolverines, and others that once thrived in the deciduous forests have survived, though in greatly reduced numbers, by retreating into wilderness areas. But it has been only in recent years that census methods have been refined enough to be able to determine the accurate status of wild animal populations.

No one ever suspected, for example, that the passenger pigeon could be annihilated. It existed in numbers beyond counting. The female passenger pigeon laid only one egg a year, but this was enough to sustain and to build the flocks before man invaded their world. Single flocks of these birds were said to contain as many as 50 million. They darkened the sky—literally blotting out the sun—as they passed overhead, nothing but birds visible from horizon to horizon. One such flock was estimated to be 50 miles wide and 250 miles long. Where these gigantic flocks nested or where they stopped to roost at night, they broke branches from trees, and their droppings, to several inches deep, whitened the ground below like snow.

In those days no one was concerned about saving the birds. Rather, they were most anxious to prevent the birds from destroying their crops, and so they combined eliminating what they considered to be a pest with getting meals for their tables. More millions of pigeons were slaughtered for sale in markets, hauled by the hundreds of tons to New York. Hogs were fattened on their carcasses. An ambitious hunter could kill as many as 10,000 birds in a day. The massacre went on year after year until at last the massive flocks were reduced to scattered groups. Even in the late 1800's, however, bills proposed to protect the pigeons were not passed because legislators believed the birds were too abundant and too prolific ever to be in danger of extermination.

More subtle, but perhaps even more damaging to the passenger pigeons, was the destruction of their habitat, for the birds were adapted to a life strictly in deciduous forests where they nested and also got their food. They could not adjust to the drastic change in their world. By the 1890's the passenger pigeon had become so scarce that it was no longer profitable to hunt the birds commercially. Further, the surviving birds were not producing enough eggs to replace themselves. A few birds were seen—and promptly shot—in the early 1900's, and then there were no more passenger pigeons.

The last living passenger pigeon, a lone female named Martha, resided in a cage in the Cincinnati Zoo where she had hatched from an egg some thirty years earlier. She died there in 1914, never having flown with a flock of her kind. She was the last of what had once been the most abundant of all birds in the world.

There were other casualties as the trees continued to tumble. The heath hen, a close relative of the prairie chicken, lived in the woodlands of North America, showing a distinct preference for oaks, and it got its food at the forest

PASSENGER PIGEON

CAROLINA PARAKEET

edges. In the early days, the heath hen was so common that it became almost a staple in the diets of some of the colonists. Dogs and cats took a toll of the ground-nesting birds, too.

Eventually protective laws were passed, but little was done to enforce them. By the early 1900's, the fewer than a hundred surviving birds were all in a preserve on Martha's Vineyard. Happily, this flock grew until it contained several thousand. Then a series of natural disasters struck. First there was a devastating fire. This was followed by an uncommonly cold winter. Suddenly the flock was down to about a hundred birds again, and in the fire-cleared land, predators could reach their mark more easily. A disease followed, and finally, in 1932, the last of the heath hens died.

Gone, too, are the Carolina parakeets that were once abundant in southern woodlands. They were killed for sport and also because the little birds made pests of themselves by eating fruit and other crops. Others were caught alive for keeping in cages. The last of the parakeets in the wild presumably died in the Florida Everglades in the early 1920's. A caged specimen kept in the Cincinnati Zoo passed away in 1918. It had been a neighbor of Martha, the last passenger pigeon. But the handsome Carolina parakeet, with its metallic green body, yellow neck, and orange to crimson crown, died with little attention being given its disappearance. Or was that really the last of the little birds? Reliable ornithologists did report seeing a small flock of Carolina parakeets as late as the 1930's in South Carolina.

The ivory-billed woodpecker, native to deciduous woods in southeastern United States, has become another victim of the cutting of the trees. The largest of all the North American woodpeckers, a striking black and white with a bright red crest and measuring more than a foot and a half long, it is believed that a few of these birds may still exist in the deep woods of Louisiana, Florida, and the interior of Cuba. Reports of sightings still are made occasionally, but none has been confirmed in recent years.

The ivory-billed woodpecker succumbed partly because of its highly specialized diet, for it fed only on the larvae or grubs of wood-boring beetles that live in the wood of dead trees, especially ash, oaks, and cypresses. These dead trees are cut along with the living in the leveling of forests. Their replacement comes too slowly to benefit the big woodpeckers, for these trees are found only in mature forests. The ivory-billed woodpecker was never common, but perhaps it could have been spared by providing it with sanctuaries. A retreat of this sort has been established in a northern Florida forest where the birds were last sighted, but most people believe the birds are now gone forever, their plight recognized too late.

When America was first settled, wild turkeys were common almost everywhere east of the Rockies and southward into Mexico. This abundance and the fact that the wild turkey was distinctly a native prompted Benjamin Franklin to propose it as the national symbol in preference to the bald eagle, not exclusively American. Benjamin Franklin did not live to see the wild turkey become almost extinct, many years before the bald eagle also became imperiled.

Wild turkeys live in woodlands where they feed on acorns, beechnuts, and similar foods. As the forests were cut, the turkeys disappeared, but they were also heavily hunted. At one time in the East, a wild turkey could be bought for as little as one cent, and some of the gobblers weighed more than 40 pounds.

By the early 1900's, wild turkeys had disappeared from about three fourths of their original range. But as the land has grown back in forests, the wild turkey population has also built up again. Wildlife biologists have aided by rearing them in suitable areas.

Another woodland species that has staged a remarkable comeback is the handsome wood duck. Reduced to rarity both as a result of habitat destruction and also hunting (for the colorful feathers and also because the duck is good to eat), the wood duck once seemed doomed. Given protection by rigidly enforced laws that prohibited hunting and then provided with nesting boxes to take the place of the trees in which they had nested, the wood duck population has now rebuilt to a presumably safe level. Reforestation of broad areas has also helped. Now the incredibly beautiful drakes in their breeding plumage again grace the American scene, an example of what can be done when there is genuine concern for wildlife.

Tales such as these could be told for literally dozens of animals that once lived in the forests of temperate regions around the world. This is where man's onslaught against the environment and its inhabitants has been greatest. Here for many centuries man has been remodeling the face of the earth to suit his needs. Some of the extinctions may have been unavoidable; others were unnecessary tragedies. Hunting animals into extinction is certainly inexcusable, but this has been done.

More sinister are those extinctions that have come about due to changes in the environment, for these same changes can also affect the well-being and even the survival of man. The most dangerous are the chemical pollutants. They may not kill immediately or directly, but the effect of the poison accumulates and pyramids. The first to be killed are the top predators in the animal community, for they become the reservoirs for dosages of poison from every animal they eat. Man, too, must be more watchful of the build-up of poisons in his own foods.

STREAMS, LAKES, AND WETLANDS

GREAT BLUE HERON—
Middle Atlantic marshland

OF ALL the water on earth, only about 2 percent is fresh, and more than half of this is permanently locked in snow or ice in polar regions or on high mountains. Streams, ponds, lakes, and wetlands thus contain less than 1 percent of the earth's waters. They form a small but very significant environment.

Compared to the seas, freshwaters are much less uniform in both physical and chemical characteristics. Great differences in the oxygen content, amount of food, temperature, light, and other factors occur not only from one body of water to another but also in the same body of water at different times of day and from season to season. Some streams flow swiftly, others are sluggish; some lakes are deep and cold, others warm and shallow. As a result of these and many other differences, the kinds of plants and animals inhabiting freshwaters are also extremely varied. Their world has been greatly abused by man, and so it is not surprising that many endangered animals are from freshwaters.

Since the beginning of time, man has chosen to live along or near streams and lakes. They have provided him with water to drink, served as transportation routes, furnished food, and increasingly important in recent years, offered places for recreation. Almost all of the food and nutrient cycles in freshwaters are intimately linked to the land. What man does to the land thus affects the waters and all life forms that live in them. The immensity of the seas has kept this effect masked until very recently, but streams, ponds, lakes, and wetlands reflect rapidly man's abuse of the land.

In the wilderness areas of the North Woods, in remote mountainous regions, in some parts of the tropics—there are still places on earth where streams run clear and lakes are untainted. But in the world's heavily populated regions, virtually no body of water has escaped the influence of man. Streams have become open sewers, and lakes are the cesspools or catch basins. In the United States, for example, every major river system is now polluted—so much so that they may soon be so lacking in oxygen that they cannot support life.

The Hudson River receives about 200 million gallons of sewage from New York City every day. Parts of the river are almost sterile of all life except bacteria. The sewage moves out into the sea as a thick, creeping carpet of sludge, and there it continues to stifle life. The condition is improving but is still unsatisfactory. The Potomac River carries almost as much pollution from the Washington metropolitan area. In fact, every city located on a river is a polluter. In the United States and in most of Europe, Asia, and Latin America, people draw drinking water from these polluted waters. It is purified with chemicals, of course, but there is a limit to how much chemical content can be tolerated. Meanwhile, the fish and other animals once abundant in these waters are being exterminated. Even the purifying chemicals become pollutants.

Hot water from power plants has recently become one of the most destructive of pollutants. In the United States, for example, more than 50 trillion gallons of superheated water are discharged into waterways every year, the amount continuing to increase at an accelerated rate as more plants use nuclear energy, producing more heat and requiring more water for cooling. The heat sometimes causes massive fish kills directly, due principally to lack of oxygen but also because it speeds the poisoning effects of the various chemical pollutants already in the water. Over a period of time the heated water brings about permanent changes in the aquatic habitat, the area near the source of the water becoming too heated to support any kind of life. Some of the spawning streams for salmon have been destroyed by heated water as have nursery grounds for young fish.

Cities are not the only polluters. Much of the damaging pollution in many areas comes from agricultural lands. Insecticides applied to crops and to forests find their way into freshwaters, killing the fish and their food animals and also the birds and other predators that feed on the fish. Chemical fertilizers can also be devastatingly damaging. They cause rampant growths of tiny plants or algae. Eventually all of the available oxygen is used by the plants, which then die due to suffocation. The fish and other animals living in the body of water also die. The decay of these plant and animal bodies further complicates the water's chemistry, making it unlivable.

Silt—consisting mostly of valuable topsoil eroded from the land—washes into streams and lakes or is carried by streams to the sea. Every day this loss amounts to more than 8,000 acres of irreplaceable soil in the United States alone. As the land becomes increasingly poor, the silt destroys living conditions in the water, choking out the growth of plants, preventing the hatching of eggs, clogging the gills of fish and other animals, and settling over the bottom in a thick, useless ooze.

Lakes and ponds go through a natural aging process. Over many years—even many centuries in the case of large lakes—their basins gradually fill, while the vegetation at the edges grows toward the center in an ever-tightening circle. This is called succession, and if it runs its full course of stages, a swamp or a marsh eventually forms in what was once the bed of a lake or pond. Finally a forest grows in the old basin. At each stage, new communities of plants and animals appear, but in natural succession, the changes all occur gradually.

When man intervenes, the aging is often hastened far beyond the usual pace. In the 1960's, for example, Lake Erie reached a "near death" stage, hurried toward its final days because its waters had become so enriched with fertilizers that the thick growths of algae used up all of the available oxygen. Fish kills were massive. But Lake Erie's condition was diagnosed soon enough to stall the lake's death. Now the lake is being rejuvenated, and perhaps a lesson has been learned. In the case of many lakes, large and small, throughout the world, the ending has not been as happy. The bodies of water die without revival and take with them all of their inhabitants.

Wetlands are important resources, too. Like sponges, they soak up water and then feed it slowly to the surrounding land. They also support special wildlife populations—frogs, toads, turtles, and snakes as well as fish and other aquatic animals. Muskrats, mink, and many other mammals thrive in wetlands, as do many kinds of birds. Ducks and geese use wetlands as nesting sites. Because waterfowl have the support of an enthusiastic following of sportsmen who have over the years paid special fees for the privilege of hunting, many wetlands have been preserved to maintain this sport. This has benefited wetlands wildlife generally.

Swamps and marshes have not escaped the pillaging by man. Many have been drained and filled for farming or for housing developments. But wetlands are not wastelands, as generally believed in the past. They are important to the ecological well-being of the lands around them, and when wetlands go, so do the many kinds of wildlife that depend on them.

In Australia, the unusual egg-laying duckbill or platypus inhabits lakes and streams—from the lowlands to mountain altitudes of 5,000 feet. Duckbills dig burrows with underwater entrances leading to above-water chambers for resting and for their nests. During the day, a platypus stays in its burrow, but at night it comes out to feed on crayfish and other small aquatic animals. In a nest in the burrow, the female lays one or two eggs and then incubates them for about two weeks until they hatch. The tiny young nurse by lapping a milky secretion that flows into hairy pockets on the female's underside. She does not have mammary glands like those of other mammals.

Until recent years, the web-footed platypus was hunted and trapped for its fur. It was not until the animals became rare that the government passed protective laws to prohibit even the taking of animals for zoo exhibits. This is wise, because none of those kept in captivity have produced young. It is safer to let the animals rebuild their population in the wild.

Only two species of alligators inhabit the earth—one in the Yangtze River in China and the other in the streams and swamps of southeastern United States. The Chinese alligator was not known to the western world until the late 1800's, though its existence was mentioned in ancient Chinese literature and apparently inspired the Chinese dragon motif. In the early 1900's, the Chinese alligator was listed as extinct, but a few animals were discovered living in a remote area about 1930. Their existence in the wild is again questionable, mainly because of habitat destruction. It is hoped that the species has been preserved in captivity, however. About 60 animals are now exhibited in zoos around the world.

The much larger American alligator was known to reach a length of 20 feet in days gone by. Alligators exceeding a length of ten feet are rare today. To call attention to their astonishing abundance, early explorers of Florida said they could walk across the rivers on their backs. Millions of alligators were killed in Florida and throughout southeastern United States. Some were killed because people understandably did not want to share living space with these big dragonlike beasts. Some were used as food, the thick tail cut into steaks. But most of the alligators were slaughtered for their hides, made into wallets, belts, shoes, purses, luggage, and other leather items. Only the bellies of the alligators could be made into leather. The thick horny plates on the back were useless. Countless baby alligators were captured and sold as pets.

Laws now prohibit the sale of baby alligators for pets. Their life in captivity was usually miserable and fateful anyhow, and often equally unpleasant for their owners. The baby "alligators" now sold as pets are caimans imported from South America, a traffic that is also being halted. Laws came to the rescue of adult alligators, too, as they appeared to be nearing extinction.

Big alligators can unquestionably be nuisances and even dangerously miscreant near habitations. They have been known to stalk dogs and even to attack human beings, and they can be especially dangerous if they are tormented or if tempted with food. These mavericks can be moved into more remote areas.

In their natural world, alligators earn their keep in a number of ways. By keeping waterways open as they make their trails through the wetlands, they prevent the water from stagnating and at the same time open pathways for the spread of gambusias or mosquitofish. Both are important aids in mosquito

ALLIGATOR

control. Also, their dens or "alligator holes" are often the only available sources of water for wildlife and livestock during the dry months of the year.

Though its range is much restricted as compared to earlier times and the total population is greatly reduced, the alligator has apparently been spared extinction. In Everglades National Park and in similar state and federal parks and reserves, the alligator is protected. Surplus populations move out into the surrounding areas, and to keep their numbers limited and controlled in some regions, hunting has been permitted again for alligators of specified sizes.

The most unusual of the crocodilians are the slim-snouted gavials that live in the rivers of India, ranging from freshwater through brackish. Museum specimens are evidence that some of these beasts reached a length of 30 feet. Twenty feet is not uncommon today. Though they are protected by law and in some regions are considered to be sacred animals, gavials are nevertheless becoming rare. Their status must be appraised regularly to assure their survival.

Sharing the Everglades with the alligator is the Everglade kite. At one time it was found as far north as the Florida panhandle, and it still ranges southward

EVERGLADE KITE

through Mexico and Cuba into northern South America. The Everglade kite is a victim of its own specialization, for it eats only the plump apple snail, using its slim, sickle-shaped bill to extract the morsels from their shells. In Florida, where its wetlands habitat has largely disappeared due to drainage, the snail has become scarce. Most of the snails have been forced to live in drainage ditches, and there they have become infested with liver flukes. The flukes in turn infest and kill the kites.

The Everglade kite appears to be doomed in the United States, though there is some hope that it may survive in parks or reserves. The chances of the bird's survival in Central and South America are greater—at least for the time being. As these areas are cleared and drained, the kite will die there, too. The only way it might be saved is by setting aside large tracts of land where both the snails and the hawks can prosper, but it will be a question of whether the land is needed more for people or for a highly specialized bird. If the lands can at the same time serve other purposes, which would surely be true in these areas, the kite can be spared.

Plume hunters once reduced both the snowy egret and the common egret to near extinction in the United States. Literally hundreds of thousands of these birds were slaughtered for decorative additions to hats and garments. It took the death of an Audubon Society warden, killed by greedy plume hunters, to focus attention on the plight of the birds. Laws were passed prohibiting the use of the plumes, and both birds are again abundant.

Fish are obviously the reigning animals in the world of water. As their habitat becomes polluted, however, they either die or must move to other areas, which is rarely possible in the case of fish. All of the fish that have become extinct over the years were not well known or prized for either food or sport, but they were distinct parts of their particular world of water. They fitted into webs of life that are now weakened or destroyed by their absence.

Of the 62 animals that became extinct in the United States between 1850 and 1870, about 14 percent were fish. But remember that the fish existed in a very small percentage of the total environment. The plight of the world of water shows even more strongly when the current list of endangered species is examined, for about 27 percent or more than a fourth of the total are fish. Of the threatened species—that is, those not in immediate danger of extinction but with their numbers declining steadily—33 percent are fish. Obviously, the fish's world plus all of the animals that share it with them is becoming a perilous place for living at a much more rapid rate than any other major habitat.

Now endangered in the United States are about a dozen species of trout, favorites with fishermen. Also on the list are three species of sturgeons, three ciscos, two bass, and many other species, mostly smaller. The Arctic grayling is rapidly diminishing in numbers, and the Atlantic salmon is no longer able to reach spawning areas in New England streams, its route either blocked by dams or by pollution. The blue pike, once one of the most important commercial fish in Lake Ontario and Lake Erie, is now near extinction.

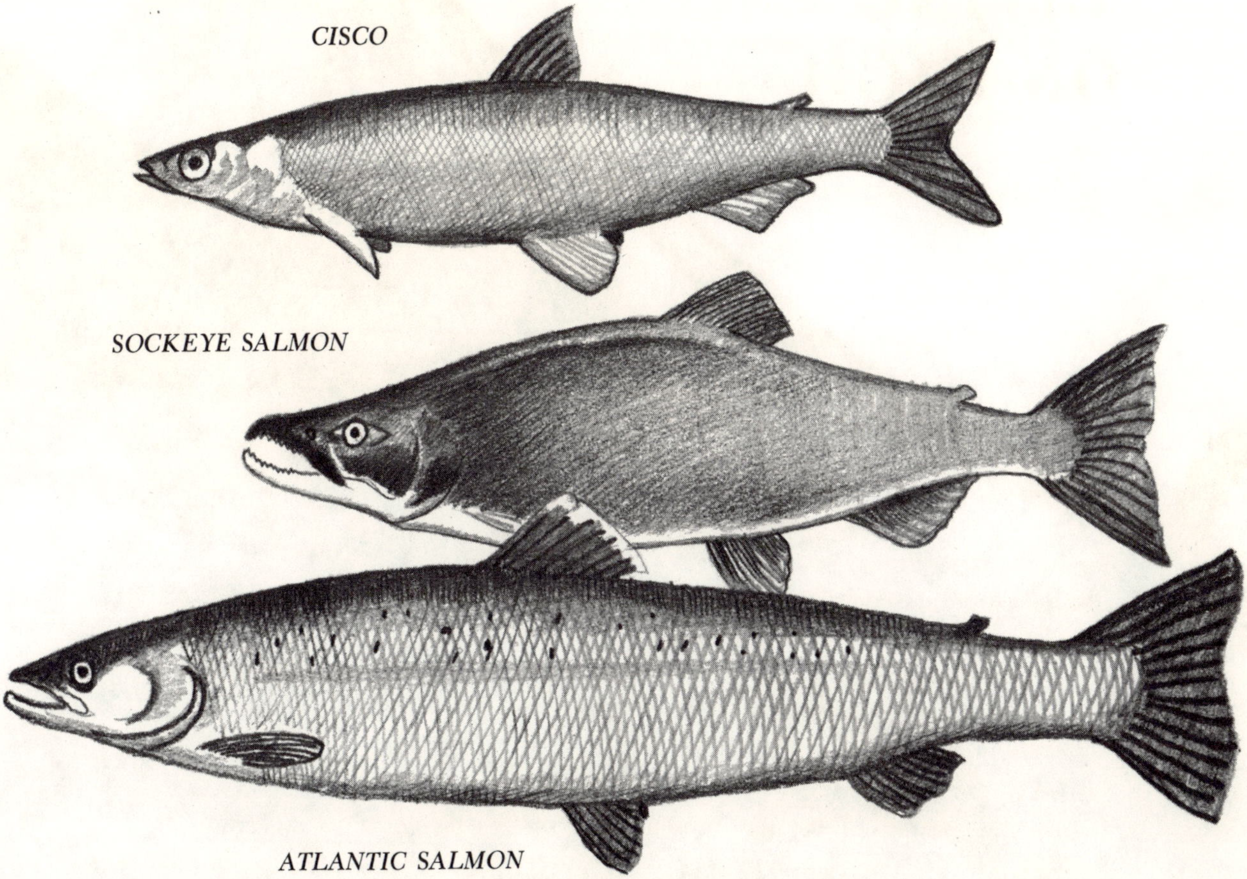

Here and there are indications that the ways of the past have been sufficiently mended to permit at least some streams and lakes to recover. Once, for example, the Adams River in British Columbia was rated as the most important of all streams for the commercial harvest of sockeye salmon. Beginning in the early 1900's, however, the various spawning areas upstream were blocked by dams or other obstructions. The sockeye salmon began to disappear. A highly valuable commercial industry was destroyed, and a species of fish was a step nearer to extinction. Canadian and American conservationists worked together to eliminate the blocked spawning sites, and the sockeye salmon have returned, demonstrating that the environment can be manipulated to favor the natural inhabitants. Man is always more aggressively interested in projects of this sort when there is a profit motive.

Freshwaters are indeed a monitor of how man treats the land. And sadly, they are also an indication now of what might happen to the land itself in the future. Fortunately, more heed is being paid to correction of mistakes made in the past. The world of tomorrow might be less hazardous for all living creatures—including man.

LIVING ON HIGH

HOARY MARMOT—
Rocky Mountains

The peaks of the earth's high mountains are perpetually capped with snow, the living conditions almost identical with those in polar regions. Below the blustery snow cap is a treeless region like the tundra. Still farther down the mountain is a zone of coniferous trees, followed by deciduous trees, and finally, depending on the latitude, a tropical forest. All of these life zones occur on only the highest mountains in the equatorial regions. The base of a mountain obviously will have the same life habitat as the area in which the mountain is located.

In the highest mountains, invasion by animals is limited not only by the cold but also by the much lower atmospheric pressure and the greatly reduced amount of oxygen. Yet there are some birds and mammals that never occur at elevations less than 10,000 feet. Growing plants need oxygen and good soil as well as above-freezing temperatures for at least a portion of the year. The steep slopes of mountains are constantly eroded, hence a rich soil does not build up. In these conditions only such hardy plants as mosses and lichens can grow. Because of these limiting factors, mountains have very distinctive populations of plants and animals.

Some kinds of animals go into the mountains in summer and then move back to the warmer, more hospitable lowlands in winter. Other animals found now only in mountains have taken refuge there as a last stronghold from persecution and execution by man. Their existence is precarious not only because of the difficult living conditions but also because their numbers are already greatly reduced. Any additional habitat changes or assaults by man might be devastating.

The Andes Mountains of South America extend the full length of the continent—from Tierra del Fuego at the southern tip northward along the west coast to Central America. The mountainous islands of the Caribbean are formed of the peaks of this chain protruding through the ocean. Surpassed in height only by the Himalayas, the Andes form an almost impenetrable barrier between the Pacific and the Atlantic shores. The highest of the mountains is Aconcagua near the Chilean border in Argentina. It towers 22,835 feet above sea level. Even the few passes in the Andes are all above 10,000 feet.

Two members of the camel family, the vicuña and the guanaco, inhabit the Andes. They are remnants of days when the family was much more abundantly represented in the Americas. The vicuña is now an endangered species, and the guanaco is becoming increasingly rare. Their nearest relatives are the domesticated llamas and alpacas, both descendants of the guanaco and both living at lower elevations. The llama is used mainly as a pack animal, and the alpaca provides both wool and meat.

Vicuñas, found only at elevations above 10,000 feet, have long been rounded up in the wild and then slaughtered to get their wool. The Incas valued the vicuña both for its meat and its wool, and they took their harvests wisely, selecting the animals carefully and then hunting the same area only once every four years. Mass slaughter did not start until the arrival of the Spaniards. Even so, the population of vicuñas in recent times was estimated to be nearly a million. The most devastating kills, reducing the vicuña population to only a few thousand animals, came in the 1950's and 1960's. The few remaining animals are protected by law, but enforcement is difficult and hazardous, the wardens regularly forced into shoot-outs with the poachers.

Also close to extinction in the wild in the Andes is the chinchilla, a squirrel-sized rodent long prized for its silky fur used to make expensive capes and coats. Though the animals are now very rare in the wild, they are raised on fur farms where the females produce two or more litters per year. Pelts from

VICUÑA

chinchillas kept in captivity are satisfactory but not as good as those from animals taken in the wild. Wise harvesting could preserve this natural resource, though some conservationists say it may be too late to save the species. Two of the three races of chinchillas are believed to be extinct already.

On the endangered list, also, is the mountain tapir, which lives at elevations of 5,000 to nearly 15,000 feet. All of the species of tapirs in the world are victims of natural predators as well as being hunted by man. Most of them do well in captivity, however, and it is hoped they may be spared extinction by preservation of breeding stocks in zoos. Unfortunately, the mountain tapir, the rarest of all tapirs and distinctive because of its thick, woolly coat, has so far not fared well in zoos. At the same time, it is becoming scarce in the wild; fewer than 2,000 animals are believed to exist today.

Other animals of the Andes may also be in trouble, as little is known about the populations of some. Among these is the long-tailed Andean cat, which lives on the high slopes of the mountains and makes its meals of vizcachas and other rodents. Geoffroy's cat, about three feet long or only slightly smaller than the Andean cat, lives at a lower elevation. It is apparently not as abundant as in previous years, but no one is certain of its exact population status.

The spectacled bear, native to the northern Andes, is one of the world's most unusual bears and has also become one of the rarest of all the animals in the world. Males are about 2½ feet tall at the shoulders and weigh about 250 pounds, sometimes more. Females are smaller. Both have yellowish-white markings that circle their eyes like spectacles. Though it is more of a vegetarian than are other bears and cannot be given the label of a marauder, the spectacled bear has nevertheless been a challenge to hunters—because it is large and because it is a bear. This continued slaughter by hunters and the steady destruction of its habitat have lowered the spectacled bear's population in the wild to a critical low point. Fortunately the bears do well in zoos, which may spare the species from extinction. The bears do give birth to young in captivity, and so it may be possible to rebuild the population.

The mountain ranges stretching southward from Alaska through Central America also harbor the remnant populations of several endangered species. One of these is the imperial woodpecker, resembling closely the similarly imperiled ivory-billed woodpecker that inhabits the woodlands of the southeastern United States. The few remaining imperial woodpeckers live in the mountains of Mexico.

One of the best known of all endangered birds is the California condor. Fewer than 50 of these giant vultures are alive today. Their 9½-foot wingspread places them among the largest of all flying birds, the span exceeded by about two inches by the closely related Andean condor. The California condor is truly a relict species or "living fossil"—that is, it existed many thousands of years ago in the days of the mammoths and the saber-toothed tigers. Even in more recent times, the California condor ranged over most of what is now the United States and southward into Central America. A hundred years ago it

CALIFORNIA CONDOR

still winged its way over most of the Pacific coast. Then suddenly its range was reduced to two small mountain counties in California where the few remaining birds live in protected sanctuaries today.

What happened to the condor? Both the destruction of its original habitat and the elimination of its food supply contributed to its steady decline in numbers. Condors are scavengers, and with the coming of man and the settling of the land, dead carcasses could no longer be found. At the same time, men with guns could not resist bringing down from the sky these almost unbelievably big birds. They also destroyed their nests and the young.

Condors never produced young in abundance. A female lays only one egg every other year. If the egg hatches, the helpless little bird is totally dependent on its mother for more than six months and then continues to stay with her for still another six months. All sorts of catastrophes can prevent the young from ever reaching maturity. Few people now expect the California condor to survive as a species until the end of this century.

Also endangered is the closely related Andean condor, an even larger bird that lives in the high Andes of northern South America. Fewer than 50 of these giant birds are believed to exist today. In the mountains of Europe, Asia, and northern Africa, the bearded vulture or lammergeyer is another large scavenger that is becoming rare over most ot its original range. Actually a kite rather than a true vulture, the lammergeyer feeds mostly on what the vultures leave behind—the bones. Lammergeyers carry large bones several hundred feet into the air and then drop them onto the rocks below to shatter them and expose the soft and edible marrow. The birds themselves have been seen soaring at altitudes exceeding five miles.

The Rocky Mountain goat of North America is not a true goat. It is related to the European chamois of the Alps. An alpine animal, it has few enemies other than man, who has steadily reduced its numbers to a population of less than 15,000. It is not endangered at the present time, though its population should be watched carefully. It now appears to be safe in the protected habitats of several national parks in Canada and the United States.

Probably the rarest of all mammals in North America is the red wolf. Once the red wolf roamed over most of central and southern United States, but with the settlement of the land, it was forced into retreat. The gray or timber wolf moved north, surviving today principally in the Canadian wilderness, but the red wolf moved southward, civilization continuing to surround it while at the same time usurping its living space. Few escaped the gun and the poisoned baits. Those remaining live mainly in Mexico, Texas, and Louisiana and possibly some in the wooded mountains of the Ozarks. The actual number in the wild is not known. A few red wolves live in zoos, which may be the best hope for the survival of the species.

Though not confined to the mountains—some of its remnant population living in the Florida Everglades and other non-mountainous regions—the puma has taken to the western mountains as a last stronghold. Known also as

ROCKY MOUNTAIN GOAT

SNOW LEOPARD

cougar, mountain lion, panther, catamount, and at least a dozen other names, the puma is the most widely distributed of all the large mammals of North America, ranging from southern Canada to the tip of South America. It has made itself at home in the tropics, deserts, and mountains, though its preference is for forested areas.

Once it occurred from coast to coast in North America, not always as a resident but at least as a transient wanderer. This giant among American cats, the largest measuring nine feet from the tip of the nose to the tip of the tail and weighing well over 200 pounds, was well known even in New England, where stories of sightings persist to this day but are not confirmed.

The puma preyed originally on whitetail deer principally, though hungry cats would never turn down even a mouse. There is a record of one individual that had a penchant for slugs, which are shell-less snails. As the land was settled and the deer disappeared, the puma turned often to what replaced the deer: man's livestock. It took only a few miscreant individuals to turn the full wrath of man against this handsome cat. Bounties were paid, and some men made it their life's work to exterminate pumas. No one knows how many were killed, but all agree that it was many thousands.

It is perhaps most remarkable that any of these cats exist today. But the puma is crafty, a tenacious species that somehow manages to cling to survival. Listed among the endangered animals, only a few thousand are believed to be alive today. It is fortunate that these big cats do well in captivity and also breed there. Many even become reasonably tame, though their tempers are likely to be easily fired as the cats get older.

Though this big cat has been an important part of the American scene since early times, little is really known about its habits in the wild. The puma shies from human beings whenever possible, often making itself scarce in situations in which other big cats become foolishly and too inquisitively bold. It is unfortunate that the puma has not done this more often.

The snow leopard that lives in the mountains of central Asia is another of the handsome big cats that is presumably safe from extinction. It could easily be threatened, however, if the fashion market were suddenly swept by a fad for the snow leopard's soft gray pelt spotted with rosettes of black. In the Himalayas, the snow leopard ranges from 5,000 to 15,000 feet, and like the puma, it keeps away from humans. But despite this wariness, snow leopards are killed and trapped regularly. Fortunately, these cats do breed in captivity, and they are exhibited in a number of zoos. Special accommodations are necessary for keeping them, however, because they are sensitive to heat.

The clouded leopard, a more abundant species that lives at lower elevations in the mountains, has a notably long tail, accounting for about half of the cat's six-foot length. The clouded leopard is another of the big cats that has a questionable safety status in the wild. The fact that it does well in zoos is the greatest assurance of its survival.

Giant pandas, sometimes placed in the same family with raccoons but generally classified with bears, are among the animals that have moved into the mountains for refuge. Once giant pandas roamed over most of eastern Asia, but they were not known to Europeans until discovered by Père David, the same French missionary for whom Père David's deer is named. The first giant panda to reach the western world was a half-grown animal delivered to the Chicago Zoo in 1937. Before 1940, eight other animals were acquired by United States zoos and still more were delivered to zoos in Europe.

The government of China then placed these handsome, unusual animals under strict protection in their mountainous retreat in southern China. There, at an elevation of 5,000 to 10,000 feet, the giant pandas live in bamboo forests. Often weighing more than 300 pounds and attaining a length of five feet, giant pandas are wholly vegetarian, eating only the shoots and tender young branches of bamboo. In captivity, the animals have produced young only in the zoo in Peking. This makes it all the more essential that the giant pandas, now rare in the wild, are not annihilated. Such a fate could befall them quickly.

In heavily populated and highly industrialized Europe, the mountains have become sanctuaries for many kinds of birds and mammals. In recent years, animals long ago exterminated in some mountain areas have been reestablished by the creation of parks and reserves where they can live in safety.

The Alpine ibex, a wild goat, once inhabited the Alps in large numbers. Since earliest times, the Alpine ibex was hunted by man, either to get trophies —the 3½-foot, sickle-shaped horns were much prized—or to get various parts of the animals used for curing ills, based on strong folk beliefs. As early as 1800, the Alpine ibex was believed to be headed for extinction. Only a few dozen animals were known to exist. These few animals were claimed by one Italian family, and under their careful, insistent protection, the population of the animals slowly increased. Establishment of large mountain parks and preserves then assured the animals of places to live. Now there are about 10,000 of these ibexes, and the species is no longer in imminent danger of extinction.

The Cretan wild goat, a very similar animal, lived in the White Mountains of Crete. When the animals were reduced to a few hundred, efforts were made also to protect them and to establish herds on other islands. But this tale does not have a happy ending. The animals were already too few in number to respond successfully, and the Cretan wild goat is now extinct.

OCEANS: THE BIG WORLD OF WATER

SPERM WHALE

SWORDFISH

To MAN, the seas have always been symbols of power and great mystery. The envelope of salty water wrapped around the earth is indeed awesome, covering 90 percent of the Southern Hemisphere and 60 percent of the Northern Hemisphere. Hidden beneath the surface are deep canyons, high mountains, broad plains, and powerful undersea currents—all in a scale making similar features on land diminutive by comparison. Mt. Everest, the highest mountain on earth, towering 29,141 feet above sea level, would be covered by water 1½ miles deep if it were dropped into the deepest part of the sea, the Marianas Trench.

Despite the immensity of the seas, which occupy more than 70 percent totally of the earth's surface, they offer remarkably uniform conditions for life. Over great expanses, the salt content is the same, and the temperature changes only gradually, differing little from day to day and from season to season. These are not conditions that result in development of new forms of life, but the individuals of an existing species commonly occur in prodigious numbers.

It is perhaps most astonishing that this vast ocean world, much larger than all of the land environments combined, is being despoiled by man. In some respects, the seas are more fragile than the land environments. The seas are safe from the plow and the ax and from fire, but man is plundering them with poisons, the most devastating way of all.

In the seas, the basic food is plankton—the microscopic plants and animals, so tiny they cannot be seen with the naked eye and yet so abundant in the great "pastures" of open water that they give it a characteristic color. The plankton feeds countless kinds of fish and other creatures. Surprisingly, the largest of all sea animals—the blue whale and the whale shark—feed directly on plankton, as do also the great schools of herring that form fisheries on which the economy of entire nations depend. Plankton feeders in turn become food for other animals in the food chains of the sea. As the tiny one-celled plants in the plankton manufacture food, they also release oxygen into the atmosphere. This accounts for as much as three fourths of the free oxygen in the air.

Already there are strong indications that the plankton pastures of the open ocean are being affected by man's pollutants. Microscopic forms of life are not included on the lists of rare, threatened, and endangered wildlife. Most people would consider such additions frivolous and ridiculous. But if the plankton pastures of the sea were destroyed, man living in areas farthest from the sea would be imperiled. A bird that stands three to six feet tall, a half-ton rhinoceros, the largest of all living lizards—these are massive animals, so spectacular in size that they are easily comprehended. But the uncountable multitudes of wee creatures in the sea actually affect man's life more directly. The web of life is complex, and the oceans are indeed vital to all life on earth.

Right now the effects of man on the sea are most noticeable in coastal areas. Most of the nearly half a million chemical compounds used by man in some manner find their way eventually into the sea. Many are poisonous to living things. In a test sample in California, for example, 99 percent of the clams collected contained DDT. The amount of DDT in individual clams was not lethal, but animals eating them would get the dosage from each clam. This is the magnification process that is killing many kinds of predators. It is estimated that over half of the 1½ million tons of DDT spread over land areas since World War II is still active—most of it now in the sea.

Another major pollutant in the seas is oil, about 600 thousand tons dumped into the oceans annually as a result of wrecks, spilling, and the flushing of bilges. At least a quarter of a million sea birds are killed every year as a result of oil. More than this number died from the spillage of one tanker a few years ago off the coast of northern Europe. Less is known about the effect of the oil on marine animals—that is, those that must survive beneath the spreading suffocating film of oil. This may be the most serious threat of all to the plankton of the open sea.

The most obviously damaged at this time are the inshore areas of the seas—the bays, inlets, and estuaries. These shallows are the most productive and the richest areas of the seas. Their food production per acre is about ten times greater than the richest croplands. Here, too, are the nurseries for young fish and shellfish. And here is where man drains, fills, settles, and pollutes.

One of the most unique marine areas in the world is Great Barrier Reef, stretching for about 1,250 miles along the coast of Australia. This is the largest living coral reef in the world, but it is now being destroyed at an alarming rate. The most direct cause is a natural one—a booming population of the crown-of-thorns starfish that feeds on the living coral. But why did the starfish suddenly increase in numbers? Man in all probability contributed indirectly.

Collectors have taken many live shells from the area, particularly the tritons that are primary predators of the starfish. In addition, many of the smaller schooling fish that also prey on the starfish were driven out of the area by man-built structures along the edge of the reef. With these natural enemies gone, the starfish rapidly increased in numbers and began consuming the living coral in such large quantities that the reef is now endangered.

At present the voracious crown-of-thorns starfish is limited to Pacific waters. Biologists are concerned about what might happen to Atlantic corals if the starfish were ever established there. This is one of the possible consequences if a sea-level canal were constructed across the Isthmus of Panama, as is often proposed. This natural block between the Atlantic and the Pacific would be eliminated. Gone would be a barrier between the roughly 4,000 Pacific-side species and the 6,000 on the Atlantic side, an isolation of three million years destroyed by man. At first, biologists say, there would be an increase in the number of species on each side. Then roughly half would become extinct as a result of competition. With his bulldozers, man would set the stage for the annihilation of perhaps as many as 5,000 species! Fortunately conservationists have prevented the building of the canal until all of these consequences can be properly appraised.

Man's conquest of the seas has been mostly as an exploiter, and one of the most sought-after of the sea's natural resources has been furs. Always high among these prizes have been the sea-dwelling seals, living by the millions in colonies along coastal areas. Until man arrived, the Guadalupe fur seal existed in peaceful abundance along the southern coast of California and southward. Its slaughter to near extinction took only a few years, literally millions of the seals killed in the period from 1800 to 1850. The seal was then believed to be extinct, but in 1948, a few animals were discovered on the rocky Guadalupe Islands off Baja California. This remnant population is now protected, though enforcement is not as good as it should be. The herd is building slowly, however, and there are now several hundred animals.

A distinct southern race, the Juan Fernández fur seal, was believed hunted to extinction in the 1880's, but like the Guadalupe fur seal, it was rediscovered in the late 1960's. Roughly 30 animals are now struggling for survival on Mása Tierra, one of three islands now included in the Juan Fernandez National Park off the coast of Chile.

GUADALUPE SEAL

Almost all of the mammals in this group, which includes the walrus and sea lions, have been persecuted, their populations spared only when it finally becomes no longer economical to harvest them. The hunting and harvesting of most species is now controlled by international agreements. Without such regulations, for example, the Pribilof or Alaskan fur seal would surely now be extinct. The uncontrolled harvests in years gone by made millionaires of many hunters and furriers, almost at the expense of the species. The population has now risen to about 2 million. The Caribbean monk seal, heavily hunted for its oil in the 1700's, exists in very limited numbers today in its native West Indies. Its close relatives, the Hawaiian or Laysan monk seal and the Mediterranean monk seal are also endangered.

Earless seals are a group that have bristly hair rather than short, soft underfur, but this has not spared them from being hunted for their oil, meat, and hides. It is most surprising that any have survived. Harp or saddleback seals, for example, have been killed by the millions, as have harbor seals, ribbon seals, and others. A population of millions of ribbon or banded seals has been reduced to about 20,000 animals. The most abundant of this group are the crab-eating seals of the Antarctic, their population now estimated to be about 5 million

WALRUS

SEA OTTER

animals. Until recent years they have been protected by their isolation, but this natural safety no longer exists. Their survival will depend on agreed upon and well-enforced international laws.

The smallest of all the marine mammals is the sea otter, a carnivore. Inhabiting the Pacific, it was discovered in 1741 by the same Russian expedition that also found Steller's sea cow and other species living in Arctic waters. To five feet long and weighing as much as 80 pounds, stockier in build than the more weasel-like otters of freshwaters, the sea otter almost never leaves the sea. It eats, mates, and even sleeps there, usually only half a mile or so offshore in the beds of kelp. At night it wraps some of the strands of kelp around its body to keep from drifting out to sea.

The sea otter's diet consists mainly of sea urchins, though it also eats some shellfish and fish. When a sea otter comes back to the surface from a dive, it floats on its back while it has its meal. Some sea urchins or shellfish can be cracked open easily in the otter's teeth. Others are too thick-shelled, and so the sea otter dives to the bottom and gets a flat stone. It puts the stone on its chest and then cracks the sea urchins by pounding them against the stone.

Soon after they were discovered, sea otters were harvested for their pelts—a thick, glossy fur that brought a high price in markets. The trade lasted for about a century and a half, but most of the sea otters had been killed during the first hundred years after their discovery. By the late 1800's they were becoming scarce, and pelts sold for as much as $2500 apiece. By 1910, the sea otter was believed to be extinct. As a precautionary step just in case some still might exist, the United States, Russia, and Japan signed an agreement making the hunting of sea otters illegal.

Two decades passed before signs of the sea otter were seen again. Soon two breeding colonies were discovered. Given immediate and full protection, the sea otters have now increased their numbers to about 50,000, and the species is safe from extinction.

The complement of the Pacific sea otter in the Atlantic was the sea mink, which was twice as large as the common freshwater mink and had reddish fur. It lived in the coastal regions from Newfoundland southward to Massachusetts, the population apparently centered along the Maine coast. Pelts of the sea mink brought a high price, and so the hunters lost no time in making their harvests. By 1870, the sea mink was extinct. Oddly, its description as a species came posthumously, more than a quarter of a century after the sea mink was no longer on earth.

Manatees and dugongs, another group of marine mammals, are totally aquatic, inhabiting warm seas, estuaries, and rivers throughout the world. They have a thick, spindle-shaped body, paddlelike front legs, and no hind legs or flippers. They use their broad, flat tail to supply power for swimming, moving their heavy body slowly through the water. Their bones are solid, like ivory. The head is rounded, the eyes small and piglike, and the lips large, flexible, and bristly. Despite their ugliness, these strange beasts are said to have contributed strongly to the mermaid myth. Another name for them is sirens.

Strictly plant eaters, these animals are known also as sea cows, and their flesh is good to eat. The giant Steller's sea cow of the Arctic was eaten into extinction within fifty years after its discovery. The dugong of the Indian Ocean is also an endangered species, as are both the Amazon and the Florida manatees. Those that escape slaughter still find survival difficult because of the destruction and pollution of their inshore habitat.

Whales, dolphins, and porpoises are the most completely aquatic of all mammals. Their body is streamlined, like a fish's, and thick layers of fat, or blubber, not only insulate the body but also provide the animals with a reserve of food. In some whales the blubber is several inches thick and accounts for a third of the animal's total weight.

Whaling was an important industry in the early days. It continued until the harvests exceeded the capacity of the whales to replace themselves. Overexploitation became inevitable when whalers adopted modern harvesting methods, taking to the sea with large factory ships around which smaller boats could work and keep the big vessel supplied. A large factory ship could process 30,000 or more large whales before having to return to port. This efficiency sounded the death knell for the whaling industry and has also brought almost all species of large whales to the brink of extinction. Among them are the bowhead or Greenland right whale, both the Atlantic and the Pacific right whales, blue whale, humpback whale, sei whale, sperm whale, finback whale, and California gray whale (said now to be making a comeback).

The smaller dolphins and porpoises have also suffered, commonly trapped and drowned in the nets of commercial fishermen. It is estimated that as many as half a million porpoises die in this manner every year. Others are victims of habitat destruction.

Sea birds are also disappearing. One of these is the short-tailed or Steller's albatross, which once winged its way over the Pacific by the millions. A giant among birds, its wings spanning seven feet, Steller's albatross became the victim of feather hunters, who killed the birds in large number where they nested on islands along the coast of Asia. The largest nesting colony was on Torishima, a volcanic island about 400 miles south of Tokyo. Eruption of the volcano in 1933 and again in 1941 almost destroyed all of the few birds that remained. In 1957, the Japanese government made the island a reserve, hoping to bring the birds back again. No one knows exactly how many of these birds are alive today, but it is probably fewer than a hundred—not enough to assure their survival. There is absolutely no justification for the annihilation of this albatross or any other bird when the objective is only to get decorative feathers.

A large, flightless sea bird, the penguin-like great auk, once existed by the millions in the North Atlantic. Standing about two feet tall, the great auk was indeed colored much like a penguin. On land it waddled along clumsily on its big webbed feet, using its short, flipperlike wings to help keep its balance. In the water, however, the great auk could swim swiftly, and it could dive deep, remaining underwater for long periods of time.

Bones found along the shores of northern seas are evidence that early man in Europe began killing the great auk. The flesh was eaten, and the fat was burned as a fuel. In northern Europe, the great auk had been exterminated for many centuries and long forgotten when it was rediscovered by explorers in the New World. There the slaughter began again, the great auk becoming a staple of the fishermen working the northern seas.

First the islands of the North Atlantic were raided. Adult birds were clubbed to death. Birds not eaten fresh were salted and put in barrels for storage aboard ships or for taking to markets in Europe. The eggs, considered a special delicacy,

AUK MEAT IN BRINE

GREAT AUK

were also collected, and the feathers were saved for stuffing mattresses. The harvesting time was summer, when the birds came to shore to nest in large colonies. Only then were they easy to get. On islands that were visited regularly, pens were built as corrals into which the birds could be driven for slaughter. There were so many millions of the birds that annihilation was inconceivable.

By 1750, however, the great auk had become too scarce to make commercial hunting practical. On a smaller but intensive scale, however, the killing was continued locally. When it was finally recognized that the great auk was nearly extinct, museums commissioned collectors to get skins for mounting and for specimen cases. On Eldey Island off Iceland, a few dozen birds came ashore every season to nest. There the collectors found them, and in a short time, the last of the great auks was on its way to a museum exhibit. If the museum scientists had been less concerned about getting specimens, the great auk might have been spared, though the great damage to the population had already occurred.

Man contributed to the annihilation of the Labrador duck, but evidence indicates that the species was on its way to extinction before man arrived. The

Labrador duck lived along the Atlantic coast of North America, apparently breeding around the Gulf of St. Lawrence but wintering as far south as Chesapeake Bay. A handsome little duck, the males were bluish black with white wing patches and white on the head and neck; the females were brownish. Never abundant, the Labrador duck often appeared in shallow coastal waters to feed on shellfish. It was a slow flier, and so it was easily shot by sportsmen and market hunters. The ducks were not good to eat, and though they sometimes appeared in markets, they were not bought. Most of the massacre occurred between 1850 and 1870, the last of the Labrador ducks apparently killed in 1875. Without the hunting, the Labrador duck might even have survived, but no one will ever know.

One of America's favorite birds, the brown pelican, is now threatened with extinction. Some authorities are convinced that the brown pelican's days on earth are numbered; others hope that, with protection, it may survive. But the pelican, which ironically is the bird for which the first sanctuary in the United States was established at Pelican Island off Florida's central east coast, has not been hunted either for food or for feathers. Occasionally a pelican will anger a fisherman by competing too successfully for a fish, but typically, the brown pelican eats only "trash" fish that have little or no interest or value either to sport or commercial fishermen.

BROWN PELICAN

What has killed the pelicans is the accumulation of poisons, mainly DDT, in the fish they eat. Each fish contains only a small amount of the poison, but when a pelican eats fish after fish, its gets the poison from each one. This builds up in the bird's body—the process known as biological magnification. The DDT breaks down into a compound called DDE that affects calcium metabolism. Eggs laid by the female pelicans are then thin-shelled, and they break before hatching. This has continued year after year, and as a result, the brown pelican's population has slumped to a low from which recovery is doubtful.

Insecticide poisoning is taking its toll of other birds of prey, particularly those that haunt coastal regions where pesticides have built up in fish and shellfish. Among these are the peregrine falcon and the bald eagle, both of which are endangered species. Like all birds of prey, they have been victims of various kinds of persecution, but the poisoning by pesticides has been most devastating to their populations.

Once the majestic bald eagle, its wing span commonly exceeding six feet, ranged widely over North America, but as today, there were two distinct races. The southern bald eagles lived in Florida and along the Gulf Coast, some occurring as far north as the Carolinas. In summer these eagles often traveled much farther north, but they returned to their southern haunts when winter winds blew. Eagles of the slightly larger northern race lived from New England and the Great Lakes northward, sometimes moving southward to find food in winter. Primarily fish eaters, the eagles look for places where the water is not frozen over. Though they are powerful birds and have keen eyesight, bald eagles are not especially active fishermen themselves. They get their meals, whenever possible, by stealing the fish caught by ospreys or smaller hawks.

Choice of the bald eagle as the national symbol of the United States did not meet the approval of everyone. One of the most outspoken of the objectors was Benjamin Franklin, who complained that the big bird's moral character and questionable courage made it a poor selection. He thought that the wild turkey would have been a much better choice, proclaiming it to be honest, courageous, and more all-American.

But Congress made the eagle the official symbol of the United States in 1782, and ever since it has graced seals, money, emblems, official documents, flagpoles, and countless other objects. This is not a new role for eagles, however. These striking raptors, of which there are more than 50 species in the world, have been symbols of leadership and courage for various peoples and governments since before the birth of Christ.

Every American sees the symbolic eagle regularly, but few Americans have ever seen a bald eagle in its natural habitat. Further, their chances of doing so are becoming less each year, for the population of eagles has been on a decline for a number of years. Pesticides are a major factor. As in the case of pelicans, the eagles get a small dose from virtually every fish they eat. Each small amount is retained in the body and builds up to a damaging proportion. Eventually the females lay weak-shelled eggs that break before hatching. In

BALD EAGLE

addition, much of their original habitat has been destroyed, and in far too many instances, the birds themselves have been slaughtered, either to get the feathers for ornaments or to eliminate them as potential threats to livestock. Young bald eagles, too, are easily mistaken for golden eagles. For this reason laws now prohibit the killing of either.

Today the bald eagle populations are centered in Florida and Alaska, with only a scattered representation in other areas. Conservationists are determined to preserve this national symbol if at all possible. Both government and private organizations have active research programs for learning more about the eagle and its plight, and both government and private lands have been set aside as preserves.

The story of the peregrine falcon may have a happy ending if present efforts are successful. One of the most widely distributed of all birds, the peregrine falcon ranges from the Arctic tundra southward through Africa and South America. It occurs also in Australia. It is perhaps best known for its bulletlike dives or "stoops" that send it earthward at a speed clocked at 170 miles per hour. It makes these dives to strike birds in midair, knocking them to the ground and then returning later to pick them up. In North America the peregrine goes also by the name of duck hawk.

As early as the 1960's the peregrine falcon had been extinct east of the Rocky Mountains in the United States as a result of accumulation of DDT in its body. Elsewhere, too, its numbers were skidding to perilous lows, putting the species on the endangered list. Since the ban of DDT, some of the places formerly occupied by the bird are monitored as livable again, and so biologists at Cornell University have launched a program to bring them back.

Working with both government and private agencies, the biologists have bred peregrines in captivity and are now releasing the birds in suitable areas from New York through New England. Birds reared in captivity and accustomed to having their meals given to them are not equipped to succeed on their own in nature. This must be learned. The release process to provide this adjustment is called "hacking." First the birds are kept in an enclosure in the area where they are to be released. When they are allowed to fly free, the enclosure and the feeding schedules are maintained. The birds can return to get their food until they have mastered hunting for themselves.

The biologists also expect the peregrines to have such a strong tie with the feeding station that it will automatically become a breeding site. This will be an aid in making checks on the banded birds from time to time.

The green turtle of the Caribbean may reach a weight of 400 pounds or more. Of all the sea turtles, it has long been rated as tops for flavor, tasting much like veal as a result of grazing on the turtle grass that grows in the warm, shallow seas. Most of the other sea turtles eat fish or other sea animals, and though they may be edible, their flesh is tougher and not as flavorful. The flesh of some may be poisonous.

Green turtles get their name from the bluish-green color of their fat, prized for making a clear-broth soup. In days gone by, the green turtle ranged as far north as Florida, where it appears only rarely and in limited numbers today. Its population center has always been the Caribbean. Columbus reported the abundance of these turtles there and also named some islands for them. His Turtle Islands later were renamed the Cayman Islands.

Millions of green turtles were harvested in the Caribbean in the early days. Long after they had disappeared from the Cayman Islands, the people living there pursued the turtles through the Caribbean. Caymanders became the most famous turtle fishermen in the world, and the green turtles were so abundant that they were compared to the bison of the Great Plains.

Like the bison, the green turtle population began to diminish—so alarmingly, in fact, that conservationists feared they would soon be extinct. Part of the problem was that the turtles were fished for by people from a number of nations and usually in international waters. Controls were difficult to establish and enforce. Also, the nests of the turtles were being destroyed and robbed of eggs by people, dogs, pigs, and wild animals. Most people held little hope for the turtle's survival under these circumstances.

An international organization was formed specifically to aid the green turtle, and a hatchery has been established on Grand Cayman Island for releasing baby turtles into the sea. Harvests of the adults are being controlled. Scientists are also using the most modern techniques to learn more about the habits of the green turtle. This includes fastening radio transmitters to the turtles' backs so that their movements through the sea can be followed. With these efforts, the green turtle will be spared extinction.

Other sea turtles—the giant marine leatherback, the Atlantic ridley, and the hawksbill—are also endangered, however. Only the loggerhead is not at present on the list of threatened species, but it too is becoming scarce.

All forms of sea life, from the smallest to the largest and including many valued commercial fishes, have been affected by man. Fortunately, it is difficult if not impossible to overfish the sea by methods now used. When a population is greatly reduced, it is no longer practical to harvest that species and attention is turned to some other species. The depleted population then rebuilds. A change in fishing techniques might alter this, of course. In the vastness of the seas, it is sometimes difficult to appraise exactly what is happening to populations. But we do know that the oceans, our largest and last frontier, are not unlimited in their resources and that they must be treated wisely for the survival of all life on earth.

GREEN TURTLE

THE TROPICS

KEEL-BILLED TOUCAN–
Central American jungle

Along the equator and immediately to the north and south are the world's jungles or evergreen rain forests. The rainfall is no less than 80 inches annually —and much more in a narrow belt near the equator. One rain forest regularly records about 400 inches and sometimes more every year. Throughout most of the tropics and subtropics, the rain comes all in one season. For the remaining months—the dry season—there may be little or no rain. At the edges of the tropics and controlled largely by the prevailing winds, the jungle grades into deciduous trees or into thorn forests. Worldwide along the coasts are dense thickets of mangroves.

The abundance of rain and the year-round warmth—the temperature ranging from a high of about 100 degrees to a low of 70—have produced a great variety of plant and animal life. In a tropical rain forest, for example, literally hundreds of species of trees may occur in a small area. A deciduous forest of the same size in temperate regions will contain only a dozen or so species; a northern coniferous forest, only one or two. When the tropical forest is mature, the tall trees form an almost solid canopy a hundred feet or more off the ground, blocking out the sunlight. For this reason, the floor of a typical tropical rain forest is bare, or nearly so, except at the edges where the sun can reach the ground. Many woody vines, or lianas, are laced through the tallest trees, which may also be filled with clumps of air plants.

Cold-blooded animals tend to reach their largest size in the tropics. Here are the largest of all the snakes, the crocodiles, the giant toads, the biggest moths and other insects. Most of the warm-blooded animals, in contrast, are smaller than their nearest relatives that live in temperate regions.

Plants grow the year around in the jungle, and so food is always plentiful. But almost all of the fruits as well as the foliage are high above the forest floor. To get them, an animal must either fly or climb. The modes of travel include every imaginable variation—from sluggish sloths that hang upside down as they creep slowly along the branches to swift, agile gibbons and other primates; from hovering hummingbirds that sip nectar in the deep-throated flowers while still on the wing to big-billed toucans and hornbills that feed on fruit; and from the tree-dwelling frogs, some of which can glide from tree to tree using parachute-like webs between the toes of their feet, to slim snakes that slip noiselessly and arrowlike from branch to branch. Many jungle animals are born in the trees and never come to the ground during their lifetime.

Though teeming with life, tropical forests are also fragile. Their most abundant resources are the rain and the warm sunshine. In an undisturbed forest, either canopied high above or covered with dense vegetation below, the fall of the rain is broken, and the sunlight is filtered before reaching the earth. But, where man has cleared the jungles for farming, this process is interrupted. Disturbed, too, is the thick litter, containing important kinds of fungi that aid in the recycling of nutrients. The soil itself is poor, for about 80 percent of the nutrients in the jungle occur either in the growing plants or in the forest litter.

Jungles are usually cleared for farming by the primitive, destructive technique of cutting and burning all of the vegetation. This is fast, but it also destroys the only reservoirs of nutrients. Crops usually grow well for one or two seasons, then the pioneer farmers must move on to new land. There they repeat their slash-and-burn devastation, leaving behind a greatly depleted soil, less rich than most deserts. When attempts are made to prolong the use of the land by adding chemical fertilizers, the heavy rains leach the nutrients from the soil before crop plants can benefit. Eventually a forest will grow again, but many centuries must pass before a multistoried jungle of mature trees reappears.

Though the tropics are rich with a variety of life, the individuals of a species are generally not as numerous as in other regions. If hunters and collectors concentrate efforts on a particular species—such as an especially highly colored parrot that is popular as a cage bird or a kind of cat that either makes a unique pet or a much-fancied pelt—the total populations of these species can be severely damaged in a short time. The native people of the jungles are not sensitive to world environmental problems. They need food—and often desperately. If a parrot brings them money that will keep their family from starving or that provides other needs, there is no question about it: the parrot is destined for a cage.

A large portion of South America lies in the tropics. This consists principally of the great basin of the Amazon River, an area that is totally almost as large as the entire United States. Based on the volume of water it carries, the Amazon is the largest river in the world, flowing eastward for 4,000 miles from its headwaters in the Peruvian Andes only about a hundred miles from the Pacific Ocean. Large seagoing ships from the Atlantic Ocean can travel inland on the Amazon for a thousand miles; smaller vessels can continue for still another thousand miles. The delta of the Amazon is 200 miles wide, a depository for silt from more than 500 tributaries that lace the jungle lowlands.

The Amazon area is one of the least developed wildernesses in the world. Only in very recent times has it felt the great impact of man and his

WOOLLY SPIDER MONKEY

technology, and now there is grave concern for the preservation of this tropical region and its inhabitants. Because of habitat destruction as well as hunting and collecting, animals of South American jungles are being added to the list of endangered species with regularity.

Several kinds of monkeys have become victims of collectors. On the brink of extinction in the wild are the woolly spider monkey, all species of uakaris, and all species of tamarins. These will soon be joined by other species if collecting is not limited and controlled.

The status of the margay and the ocelot, two small, handsome cats of the tropics, is questionable because their populations are difficult to census. Jaguars have been persecuted by hunting and forced to retreat from many areas where they were once common. The small-eared dog, never abundant or well known, is still another of the mammals that are now rare in the South American tropics. Certainly the killing and collecting of all of these animals should not be permitted.

After collecting alligators in the United States was banned, caimans from South America were substituted for sale as pets. Because their stomach is plated like their back, much less of a caiman's hide is usable, yet a goodly number are still killed also for leather.

Baird's tapir, which lives along waterways in the lowlands, is another of the endangered mammals of South and Central America. Shy and never abundant,

tapirs have long been a favorite prey of the jaguars, caimans, and man. But none of these predators made as severe inroads into the tapir population as has the continued destruction of its habitat. Once distributed widely, ranging even into North America, Baird's tapir long ago was forced to retreat into the tropics as a last stronghold, but now this world is also crumbling. It may be spared only because it thrives in zoos, though the establishment of national parks in its range is also promising.

The many colorful birds of the tropics have not been spared. Now rare or endangered are some of the jewel-like hummingbirds, found nowhere else in the world. The red-billed curassow and several tinamous and orioles are among the endangered species. Despite the heavy inroads made already, the jungles of South America are still the least exploited of the world's tropics, and now that man understands what effect he has on environments, perhaps more of this natural world treasure can be spared.

Africa's tropical forests are much smaller in total area than South America's, but the problems are amplified because of man's influence in the region for so many more years. The forested region is located on the tropical west coast.

In Africa as in South America, a number of kinds of monkeys and other primates are being taken in too great a quantity for the safety of the species. The attractive colobus monkeys are already on the list of endangered species. Gorillas, the largest and most powerful of all the primates, are rapidly becoming rare. Probably fewer than 500 of those that dwell in the forests of mountain regions are alive today.

Okapis are jungle-dwelling, short-necked relatives of giraffes. Only about five feet tall, they have skin-covered, knoblike horns as giraffes do. Most of their body is a solid color, ranging from dark brown to roan, but both the front and the hind legs are striped with white, like the striping on a zebra. It is most strange that these rather large animals were not discovered until after 1900,

though early explorers made hinting references to the existence of such animals. Efforts to capture them for rearing in captivity failed until the 1930's. None of those kept in captivity gave birth to young until after 1950. Since then, more than a dozen have been born in zoos, and it now seems that the okapi is assured of survival, though the number of animals living in the wild continues to dwindle because their habitat is greatly limited. Even more rare are duikers, short-legged, forest-dwelling antelopes.

This is the home, too, of the pygmy hippopotamus. The pygmy hippo has never been abundant. It stands no taller than a domestic pig and weighs only about 500 pounds; by comparison, the common hippo may weigh as much as four tons! Also, the pygmy's snout is not as broad proportionately, and while it feeds in or near water, it always takes refuge in the dense forest rather than in the water as the common hippo does. Because it is shy and is active only at night, the pygmy hippo is rarely seen even where it is known to exist. All evidence, however, points to its being greatly reduced in numbers, due probably mostly to hunting. It has been preserved in zoos, where it mates and produces young readily.

Asia's tropics include the most heavily populated areas in the world, yet there are still dense jungles in the lowlands of India, Burma, Cambodia, and South Vietnam. Here more than in any other tropical area, man has long dominated the habitat, and in man's constant struggle to exist, pressured mostly by competition with other men, wild animals have had to give way. Starving people cannot be sympathetic to sparing wild animals for esthetic reasons. Man's own survival is more basic and personal. In their work to save wildlife, even the most extreme conservationists have learned that in such cases they must first attend to the welfare of man.

Many animals with wide ranges in years past now live in a few remaining wilderness regions. Lions, for example, once roamed over most of India but now are confined to the hot, humid Gir Forest. Fewer than 300 surviving lions are protected there.

Tigers originally ranged widely over Asia, too. Like most other big cats, tigers and man have long been in conflict, each at times playing the role of the hunter. But with his gun, man has the longest, most lethal reach. Within the last twenty-five years, the total number of tigers, including those of cooler northern regions as well as those of tropical lowlands, has been reduced from an estimated 40,000 animals to fewer than 2,000. To some degree, at least, this killing has been justified. Tigers kill cattle by the thousands and have also taken a toll of people annually—more than 500 per year until recently and as many as 1,000 per year in days gone by. Man has retaliated by virtually exterminating the tiger, the hunts organized as much for sport as to get rid of a fearsome beast. Those few that remain must be given protection.

Leopards, which also live in the jungles of Asia, have fared surprisingly well. The biggest danger is the periodic upsurge in popularity of their pelts for coats, capes, and other apparel. Only one such craze would be enough to trigger the annihilation of these magnificent beasts.

The Indian rhinoceros is definitely an endangered species. Probably no more than 600 of these one-horned, leathery-skinned animals exist today. Weighing more than two tons and growing to 14 feet long, they are confined to preserves that have been set aside for them. Their only enemies are the diseases they contract easily from domestic cattle, plus predation from the big tigers and from man, who has hunted them for sport and also to get their horns. The horns are ground up to make a powder believed to have magical properties.

Tapirs are also becoming scarce—because of predators and hunting by man and also because of the continued destruction of their jungle habitat. Regular checks must be made on their population to make certain they are not being overkilled. The largest of the grazing animals in the tropics of Indonesia are the seledangs or gaurs, wild cattle that stand as much as six feet tall at their shoulders and may weigh as much as a ton. Only a few scattered herds remain, the population lowered greatly by diseases contracted from domestic cattle. Seledangs have never been successfully domesticated, and they do not do well in zoos. Their loss would mean the disappearance of a gene pool that might possibly contribute significantly to domestic breeds. The arna or water buffalo, a smaller species living in Indochina, has been domesticated for use as a draft animal and is not endangered.

Chitals, muntjacs, sambars, and several other species of wild deer have long been hunted heavily; still they exist in surprising numbers. Like deer in other parts of the world, however, their populations could slump rapidly. Those concerned with the preservation of wildlife must be watchful of such a trend.

Large numbers of monkeys and other primates are captured for sale as pets or for use in research. Rhesus monkeys, one of a number of species in the macaque group, have served medical research importantly because of their susceptibility to many diseases of man. Rhesus monkeys breed readily in captivity, hence they are not at present in danger of extinction as a species. Their numbers in the wild are dwindling rapidly, however. As a group, the macaques spend most of their time on the ground, and they eat a wide variety of foods. Langurs, another large group of Asiatic monkeys, are leaf eaters that live high in the trees, seldom coming to the ground. Langurs do less well in captivity, hence are not as commonly caught for exhibition. The douc langur is nevertheless an endangered species, more a result of the war in Vietnam than a direct assault against the monkeys themselves.

The forests of Australia, covering less than 5 percent of the continent, have also been ravaged by hunters and collectors. Parakeets and parrots are becoming scarce, several species now among the endangered. And the bizarre laughing kookaburra, a large forest-dwelling kingfisher, is rare, its precise status not known. Several species of bandicoots and wallabies are also endangered, as is one of the most famous of all Australian animals, the teddybear-like koala. Koalas, which feed only on the leaves of eucalyptus trees, are almost totally exterminated in the wild, victims mainly of pelt hunters. There are a few captive specimens in Australia and also in California zoos. In both places, supplying the animals with eucalyptus leaves is not a problem. Fortunately, the koalas do breed in captivity, for the greatest hope for preserving the species now rests with the zoo animals.

Sprawled around the world along the muddy shores of the tropics and subtropics is a scrubby forest of mangroves, the tangled thickets of these gregarious trees walling tidal creeks and rivers for as far inland as the waters are quiet and salty. Only sandy beaches are too sterile for an invasion of mangroves. The most picturesque of the several kinds are the red mangroves, their arched prop roots giving the trees the appearance of stepping out to sea. At high tide, their lowest branches sweep the water. At low tide, the roots, their total span often as much as three times greater than the height of the tree, are exposed—clustered with oysters and crawling with crabs.

From their profusion of pale yellow flowers, the mangrove trees produce small, leathery, brown fruit. The fruit germinate while still attached to the tree, becoming long, slim, cigarlike seedlings that continue to draw their nourishment from the parent tree. When at last they are ripe, usually two or

KOALA

three months later, the slightest breeze or a slapping wave jars them loose. Arrowlike, the seedlings plunge into the mud below at low tide and immediately send out anchor roots and leafy shoots. If the tide is high, the podlike seedlings float away. Soon their root ends become waterlogged and sink. The leaf ends remain up so that the pods float vertically through the water. Weeks or even months may pass before the currents carry the seedlings into shallows. As soon as the root end of a seedling touches soil, it springs into life. Roots push rapidly into the soil, providing anchorage, and a tuft of little leaves soon marks the top.

Joined by others carried by the same currents, the sprouted seedlings become a little island of trees. Leaves, sticks, and sediment become caught in the root tangles, and an oozy land builds slowly beneath the amphibious trees. In this way, the shore of the mainland is extended into the sea, and new islands are built in the offshore shallows.

Mangroves are an important part of tropical and subtropical shores. The waters of the mangrove coast are the nursery grounds for many species of fish. When the mangroves are leveled for settlements and the swamp lands filled, many creatures from the sea have lost the place where they spend their young lives. In these hot, silent green jungles, too, are the rookeries of many kinds of birds. To man, mangroves are waste areas, but in nature, they are vital to the life of the sea and also the shore. Wherever the mangroves are destroyed, the life in the waters they fringe is impoverished.

Along the mangrove shores and the islands at the southern tip of Florida—mostly, by good fortune, within the confines of the Everglades National Park—the American crocodile clings precariously to its existence. The few surviving here are part of a population of perhaps no more than two thousand totally occurring in the West Indies, Mexico, Central and South America. The American crocodile differs from the American alligator in having a much narrower snout. Further, in all crocodiles (some of the more than a dozen species do have broad snouts), the very large fourth tooth on each side in the lower jaw fits into a notch in the side of the upper jaw, and it is clearly visible when the jaws are closed. In alligators, this tooth cannot be seen when the jaws are closed. American crocodiles live in brackish and salt waters, hence intermingle with the American alligator only briefly and by accident. Occasionally a crocodile is seen swimming in the sea many miles from shore.

The American crocodile—and virtually all other crocodiles around the world, except in remote or unsettled regions—has been pushed to near extinction for two familiar reasons: destruction of its habitat and hunting. Crocodile hides have long been valued as a source of highly durable leather for making belts, shoes, purses, and similar items. In the United States, the sale of products made from endangered species is now illegal. Enforcement of this law is difficult, but it is made much easier if people simply refuse to buy. Then the result for the American crocodile, as well as its companions around the world, will be a steady increase in numbers.

GANNET—Galápagos Islands

ISLANDS: ISOLATED BUT UNSAFE

CONTINENTAL islands are close to major land masses. They share with them the same kinds of plants and animals and are formed of rocks and soils of the same origin. They are separated from the mainland by a relatively narrow body of water. In contrast, oceanic islands are far from the mainland. Some are volcanic in origin; others are formed of coral. Because of their remoteness and isolation, their plant and animal life is distinctive.

Animals reached their oceanic island homes originally by swimming or flying or by floating to the island on debris. Winds and stormy seas have been responsible for some introductions, and in recent times, man has introduced many plants and animals to oceanic islands. The native populations of oceanic islands have been isolated from their mainland relatives for countless generations and may now resemble them only superficially.

Darwin's finches of the Galápagos Islands, as one example, have departed greatly from the basic finch type in becoming adapted to fit every type of habitat on the islands. Some have long, slim bills for sipping nectar from deep-throated flowers; others have heavy bills for probing into bark and rotten wood for insects and their larvae; still others have sharp hawklike bills, and so on. Yet all of these finches are descendants of an ancestor with a thick seed-crushing bill like all other finches in the world. The specializations of Darwin's finches equipped those living on the Galápagos Islands for occupying niches that on the mainland are inhabited by birds of entirely different families.

Remoteness from competitors is a protection for island-dwelling animals until man arrives. Then the lack of places to which they can escape makes most island animals highly vulnerable to man's assaults. And if man does not exterminate the animals himself, his domestic animals or introduced wild animals give the native species competition for food and living space while in some cases preying on them directly. Some of the most tragic examples of man-caused exterminations have occurred on islands.

DODO

Nearly everyone knows the expression "as dead as a dodo." Fewer people know how the expression came about or what the dodo was. The dodo was a turkey-sized flightless bird that lived on Mauritius Island about 600 miles east of Madagascar in the Indian Ocean. There it had no enemies until Dutch explorers discovered the island in 1598. The big dodos were good to eat, and they were easily killed with clubs. Over the next fifty years, Mauritius Island became a regular stop for sailing vessels, their purpose to stock their larders with dodos. Dogs, cats, and pigs brought to the island by European settlers ate the big, fat birds, too, and they also destroyed the nests and eggs. In less than a hundred years, the dodo was exterminated.

Would our lives be different if the dodo were still alive? We really do not know. The dodo might have been domesticated and be as important to us now as are chickens, turkeys, and ducks. Or they might have contributed to producing a hybrid fowl. Sadly, we will never know. It is especially unfortunate when a species such as this is exterminated, for there was no good reason for the continued slaughter of the birds into oblivion.

If the Mauritius kestrel does still exist, it is making its last stand on Mauritius Island today. At the last count, there were fewer than half a dozen of these birds. Mauritius kestrels nested in trees, and almost all of the trees on Mauritius Island have been cut down. Also, monkeys were introduced to the island, and they robbed the kestrel nests of their eggs. Shrubs that were brought in added to the difficulty, for they provided a protective cover for the escape of lizards from the kestrels. The lizards were the principal food of the kestrels. If the Mauritius kestrel does become extinct, which now seems to be a certainty, it will simply add another name to a list, for the Mauritius pink pigeon and the pigeon hollandais disappeared from the island long ago. Still other inhabitants of this small island on the endangered list are the Mauritius cuckoo-shrike and the Réunion cuckoo-shrike, both victims mainly of habitat destruction.

The aye-aye, another island dweller, is now near extinction on Madagascar, about 250 miles east of the southern tip of Africa. Madagascar, fourth largest of the world's islands, is the home of a number of unusual animals, but the aye-aye is one of the strangest. A primitive primate that was long classified as a rodent, the cat-sized aye-aye has a long bushy tail and thick grayish fur. Its hind legs are much longer than the front legs, its ears are large and rounded, and its eyes are owlish. Most unusual are the aye-aye's slim, bony fingers. The especially long third finger is believed to be used like a probe for picking insect larvae from holes in wood or for scooping out the edible, pithy cores of stems.

After the discovery of the aye-aye about 1780, many attempts were made to keep it in captivity, but none were successful. Exactly why the animals have steadily dwindled in numbers in the wild is not clearly understood. They were not hunted, nor were they collected in large numbers. The natives considered the aye-ayes sacred and did not bother them. Only the continued destruction of their habitat can be blamed for the aye-aye's low population today. Why save the aye-aye? Well, why not? It is a harmless, unobtrusive little animal that shares a common ancestor with man. It could conceivably contribute at some future time to man's understanding of man. Nocturnal, solitary, and extremely delicate, the attractive and appealing little aye-aye now seems destined for extinction.

On Madagascar, too, several species of lemurs, also primitive primates, are nearing extinction, and the closely related lorises are becoming rare. Again, the world would lose links with the past if these animals disappeared. The Madagascar tortoise is also an endangered species. Fewer than a dozen are now alive, making this tortoise one of the rarest of all animals. Two larger tortoises became extinct earlier, one of them as a result of man, the other by natural processes.

Another of Madagascar's unusual fauna was the giant, flightless elephant bird, which weighed as much as half a ton. Its 50-pound eggs were the largest single cells known in the animal kingdom. Several have been found in recent years well preserved in the swampy land. The elephant bird had an extremely bulky body; it stood less than ten feet tall. This behemoth among birds existed until at least the 1600's. Man apparently had only indirect responsibility for its disappearance, which was due primarily to natural changes in its habitat. But whatever the cause of this bird's extinction, gone was one of the world's most spectacular creatures that many believe inspired the legendary Roc told about by Sinbad the Sailor in the Arabian classic *The Thousand and One Nights*.

On a small island off the east coast of Madagascar, a ground-nesting coucal, which belongs to the cuckoo family, survived until about 1930. It became extinct as its habitat was destroyed. Still other species teeter on the

TUATARA

brink of extinction today on Madagascar, the destruction of habitat their principal threat. Among these are the thick-billed cuckoo and Soumagne's owl, both forest dwellers. Both may now be extinct, as they have not been observed for several years. The Madagascar serpent eagle, Anjouan Island sparrow hawk, Madagascar teal, long-tailed ground roller, Alaotra grebe—these are still other species that have become very rare, endangered, or in limbo.

On New Zealand and the smaller islands nearby are many unusual animals found nowhere else in the world. They evolved here as a result of the long isolation of these lands from continental masses. Here, for example, are the most primitive of all reptiles, the sphenodons or tuataras. All other members of their group disappeared millions of years ago. About two feet long, a tuatara looks much like a stout-bodied lizard. Its unique feature is a "third eye" in the middle of its head. In younger animals, this "eye" does indeed have a transparent cover and presumably functions well enough to distinguish light from dark.

Tuataras were once found over most of the main islands of New Zealand, but they disappeared. Some say they were destroyed by the pigs introduced to the islands by man. Others attribute their disappearance to changes in climate and habitat. Today, however, tuataras exist only on the dozen or so small islets in the strait between New Zealand's two main islands. At present, they are protected by strictly enforced laws. Tuataras will presumably not be in danger unless some capricious future governmental group elects to ignore the old laws and to convert the little islands to some use for man. This would be unfortunate, for the tuatara is one of science's treasured "living fossils."

New Zealand earlier lost one of the world's biggest birds, the moa. Until the arrival of man with his pigs, dogs, and cats, New Zealand had more different kinds of flightless birds than any place on earth. They had virtually no enemies.

The first men to arrive on New Zealand were the Maoris from Polynesia. This was about 1300. By the time Dutch explorers reached New Zealand in the mid-1600's, the Maoris and their animals had already eliminated a number of native birds and other animals. Over the years, the Europeans continued to hear stories about an extremely large flightless bird that had once been abundant. Some of the Maoris claimed that the birds could still be found in the remote areas of the islands, but no European ever saw a living moa.

The searches for the moas did result in finding their bones. The largest of the moas stood 12 feet tall, almost twice as tall as an ostrich. The smallest was no bigger than a chicken. These were the extremes of the more than 20 species of moas that once lived on New Zealand. Some of the bones were of fossil forms that existed as long as 30 million years ago.

The last living moa was apparently killed about 1800, but no one is really sure of this. The Maoris seemed to enjoy telling tales without great regard for accuracy. It seems evident that the birds became scarce after 1500, having served the Maoris importantly as food until that time and becoming increasingly scarce afterward.

TAKAHE

KIWI

Another flightless bird that still exists in New Zealand is the takahe or notornis, but it is nearing the line of extinction. A turkey-sized gallinule with bright greenish-blue plumage and a red bill, the takahe was believed to have become extinct about 1850, but in 1948, a hundred of the birds were discovered in a remote valley in the mountains. With rigid protection by the government, these birds may continue to increase in numbers.

Kiwis, the chubby chickenlike birds that lay pound-sized eggs (a fourth the female's body weight), have become a symbol of New Zealand. Kiwis, which are close relatives of the moas, live in underground burrows. They have poor eyesight, but their sense of smell is exceptionally keen for finding the worms or other small crawling creatures on which they feed. Unlike any other bird, their nostrils open at the tip of their bill, which is long. Bristles around the base of the bill serve as sensitive "feelers." While they were once heavily hunted and are no longer common anywhere, the kiwis are now protected by law and appear to be safe from annihilation.

As on many other islands throughout the world, the animals introduced to New Zealand by man are often more detrimental to the native populations than is man himself. Rats, dogs, cats, pigs—these are among the worst offenders, often becoming feral and then competing even more directly with the wildlife for food and living space. In these contests, the native species are generally the losers.

The Tasmanian wolf, a collie-sized marsupial carnivore, the rear of its body striped vertically with broad dark-brown bands, was long ago exterminated on the Australian continent where it was once common. Now it is found only on Tasmania, an island off the southeastern coast of Australia, and it is restricted there to remote mountainous regions. Even in this retreat, the Tasmanian wolf has not been safe. Because of persistent complaints that it preyed on livestock, the government paid bounties for the killing of Tasmanian wolves. This all-out eradication program combined with the elimination of much of its natural habitat brought the Tasmanian wolf to near extinction. Or perhaps the Tasmanian wolf does not exist today. No one is really certain, for none has been seen in recent years. A large reserve was set aside for this largest of all the marsupial carnivores more than a decade ago, but it has no known occupants so far. The Tasmanian devil, a badger-sized, big-headed marsupial carnivore, appears to be holding its own at this time. Like the Tasmanian wolf, however, it was once more wide ranging and was long ago driven out or exterminated from settled areas.

North of Australia are the islands of New Guinea, the Celebes, Borneo, Java, and Sumatra. These are among the largest of literally thousands of islands in the South Pacific. Several of these islands are the homes of familiar animals whose existence is now in peril.

In years gone by, as one example, orangutans roamed over much of southeastern Asia and were found also on many islands off the mainland. Their total population today is fewer than 5,000, some of the orangs living in the jungles of Sumatra but most of them confined to Borneo. Unfortunately, orangutans are still being collected for sale to exhibitors. The usual method is to find a mother with a baby. The mother is shot, and the young is then easily taken captive. More than half of these babies die before they reach their destination. Almost a hundred zoos now exhibit orangutans, and since the animals do breed readily in captivity, there is no need for continued depletion of those living in the wild. There are laws controlling the capture and sale of orangutans, but they have not been enforced. Orangutans need protection if the species is to survive in the wild.

On Komodo and three nearby islands off the coast of Indonesia lives the world's largest lizard, the Komodo dragon. Measuring eight feet long and weighing as much as 300 pounds, the Komodo dragon makes its meals mainly of wild pigs and deer, though it will also eat smaller animals or may feed on carrion. No other animals on the island are large enough to prey on this giant lizard, which has been protected by law since 1926.

But the law protecting the Komodo dragon does not include protection of the big lizard's food supply. Human hunters also kill the pigs and deer for food, and so the Komodo dragon has literally been starved into scarcity. It could easily become extinct with man only the indirect cause. Only about 300 Komodo

SUMATRAN RHINOCEROS

dragons now exist in the wild. Perhaps as many as two dozen are exhibited in zoos, but since they have so far not bred in captivity, preservation of the wild stock is important for saving the species. This is another of the paradoxes of wildlife conservation, for to save the lizard, we must first feed the people who compete with the lizard for its natural food.

Two other endangered species of the area are the two-horned Sumatran rhinoceros, with only about a hundred animals existing today, and the single-horned (females may have no horns) Javan rhinoceros, with fewer than 25 animals in existence. Both are smaller than the mainland Indian rhinoceros, and both once lived on the continent but were hunted into extinction.

The rhinos have been killed primarily to supply markets that attribute magical and medicinal powers to their pulverized horns and also to other parts of their bodies. A rhino's horn can be sold for as much as a thousand dollars so that even in sanctuaries these animals must be protected by constant patrols. And in some places the hunting continues, with handsome pay-offs to law enforcement people.

Monkey-eating eagles, occurring now only on Mindanao in the Philippines, are near extinction, only about 50 of the big birds now living in the wild. Monkey-eating eagles have a three-foot wingspread, and their plumage consists of big patches of white, reddish-brown, and black feathers. The most distinctive feature is the unkempt, frazzled crest. In addition to eating monkeys, the eagle preys on hornbills and other large birds. In turn, it is hunted by the natives and either killed or trapped alive for sale. More than a dozen of these rare eagles are now exhibited in zoos, but none have bred in captivity. Some conservationists believe that the hunting of monkey eagles for zoos has been a major factor in the reduction of the big bird's population to its current critical low. The eagle's boldness, even to attacking dogs near dwellings, has not endeared it to the natives. But this is nothing new. Nowhere in the world have large predatory birds escaped persecution. Even when they really pose no direct competitive threat to his welfare, man seems unable to refrain from killing big birds of prey.

On the nearby island of Mindoro, the tamarao is nearly extinct, with probably no more than a hundred of these wild cattle now in existence. Once it ranged over most of the island, from the lowlands to the mountain slopes. Now it is confined to the bamboo thickets high on the slopes. Though it is illegal to hunt the tamarao, there is little or no enforcement of the law. Modern hunters even make their kills from airplanes and helicopters. A small but inadequate reserve has been established.

In the Celebes, between Borneo and New Guinea, the similar but slightly smaller anoa has also become extremely rare. Its decline in numbers and retreat into the most remote areas is due both to excessive hunting and to destruction of its habitat.

Another endangered inhabitant of the Celebes is the black ape. Closely related to the macaques but resembling baboons because of its long snout, the black ape wears a tuft on its crown and has only a stub of a tail. It is a victim both of habitat destruction and capture for exhibits.

HAWAIIAN GOOSE

Farther east, in the Hawaiian Islands, more than 20 species of birds are rare or endangered. Their island homes are no longer peaceful sanctuaries—not since the arrival of man with his pets and livestock. The best known of these birds is probably the nene or Hawaiian goose, which inhabits the islands of Hawaii and Maui. When the Polynesians settled on the islands years ago, they began immediately to hunt birds for food and also to get colorful feathers for decorations. They hunted about two dozen species into extinction, but they ignored the nene, a drab, unattractive bird that nested in the lava fields.

It was the white settlers who recognized the nene as a big gray goose—and good to eat. Also, for a month each year, during its annual molt, the big goose was flightless, hence easily killed with a club. They began slaughtering geese not only for their own use but also for export, sending them to the United States by the thousands of barrels. By 1850, the nene had become scarce. Hunting was limited to only part of the year, but this only prolonged the butchery. By 1900, the great flocks of nenes were gone. The mongooses, cats, dogs, rats, and pigs—all brought to the islands by man—had also contributed to the slaughter, killing adults and young and destroying nests and eggs.

The number of nenes dwindled until finally there were fewer than a dozen birds alive in the wild. By good fortune, however, one man recognized the plight of the species and built up a flock of more than 50 birds. A tsunami destroyed this flock in 1946, but some of them had earlier escaped and thus survived the tidal wave's destruction. Finally recognizing that the nene was near extinction, the government then began working to help save the bird. Preserves were set aside on both Hawaii and Maui, where the birds are now protected from man and from most predators. An intensive breeding program was also initiated to produce birds for release in the wild. Now there are hundreds of nenes again, the number increasing every year.

One of the nene's natural predators is the Hawaiian hawk, but it never took enough of the geese to damage the total population. The hawk itself is now an endangered species, with only about a hundred birds remaining. Still more rare, with a total population of two dozen, is the Hawaiian crow. Several species of Hawaiian honeycreepers are also near extinction, one represented by a population of fewer than a dozen birds. The Hawaiian Islands have demonstrated how devastatingly disruptive man can be to the unique life of a region in a short span of time. About two dozen species of birds have already become extinct on the islands. In no other place on earth has the tragedy been greater than on these tropical "paradise" islands.

The effect of man has been felt, too, on the oceanic Galápagos Islands off the southwestern coast of South America. It was the distinct animal population of these islands that contributed to Darwin's formulative thinking in arriving at his theory on the origin of species. But others who visited the islands were more concerned about their stomachs. Here lived the giant Galápagos tortoise, some weighing 400 pounds. In the early days, these big tortoises basked on the rocky islands by the hundreds of thousands. The islands were, in fact, named for the

GALÁPAGOS TORTOISE

tortoises, for the Spanish word *galápago* means tortoise. Explorers and whalers discovered that the tortoises made excellent eating. Equally important in those days before refrigeration, the big animals could be kept alive in a ship's hold for weeks without needing either food or water. They provided the ships with a steady supply of fresh meat.

It is estimated that about 10 million Galápagos tortoises were hauled off the islands for food. Feral dogs and pigs, introduced to the islands by man, did their share of damage to the young tortoises and to the nests containing eggs. Finally, protective laws were passed to rescue the species from extinction. Today an estimated 2,000 tortoises live on the islands.

Two kinds of large lizards also inhabit the Galápagos Islands—the marine iguana and the land iguana. The marine iguana, which may be as much as four feet long, has a flat tail used to swim in the sea where the lizard forages on seaweeds. The slightly smaller, more ill-tempered land iguana lives in the dry interior of the islands where it feeds on the spiny pads and the fruits of cacti, obtaining its water from these succulent plants. On some islands, the iguanas are still abundant. On others they have been exterminated by man, who relished meals made of the iguana's thick, meaty tail, or by his various animal companions. Goats, cattle and horses ate the vegetation. Rats, dogs, cats, and pigs preyed directly on the iguanas and their eggs.

Similarly, the Galápagos penguin has been reduced in numbers to only a few thousand birds, and only about two hundred Galápagos hawks now exist. Also rare is the flightless Galápagos cormorant. Until the arrival of man, all of these birds were safe. Only those living on the smaller islands, unoccupied by man

KEY DEER

and his beasts, have been spared. Ecuador is now working with conservationists to try to restore the native populations of the islands, but it is difficult to amend mistakes that have been continued over centuries. Further, on the islands now populated by people, the competition for living space grows steadily.

All of the deer native to North America have prospered except one species—the Key deer, a subspecies of the whitetail deer. A few years ago this tiny deer, which stands only about 2½ feet tall at the shoulders, was nearly extinct. The deer inhabited the Florida Keys, a steppingstone-like chain of islands off the state's southern tip, and they had disappeared from all of these islands except Big Pine Key, one of the largest islands near the end of the chain. Over the years, a refuge of nearly 7,000 acres was established for the deer. Several hundred Key deer now live in this sanctuary, the species apparently having been saved from extinction.

The islands of the Caribbean compare with Hawaii in the devastation of native animals that has occurred in the short time since the discovery of America. Much of the damage here has also been brought about by the rats, pigs, cats, dogs, and other animals introduced by man. Goats have ravaged the landscape. But probably the most damaging of all has been the mongoose. It was introduced to the islands to help get rid of native rodents that sometimes plagued the crops, and the mongoose did perform this task admirably. Then, its numbers greatly increased, it began filling its stomach on almost every other kind of animal that moved. Few of the native species now remain.

On Cuba, one of the two species of shrewlike solenodons is now extinct, and the other is very rare. Two hutias still exist but are endangered; a third species became extinct during the 1800's. Several species of rats and two species of bats are now extinct. Among the now rare birds are the Cuban ivory-billed woodpecker, Cuban sandhill crane, Cuban sharp-shinned hawk, Cuban tree duck, Zapata rail, and Zapata sparrow. In the famed Zapata Swamp, too, lives the Cuban crocodile—only a few hundred surviving, ranking the Cuban crocodile among the world's endangered species.

On nearby Jamaica, the Jamaican hutia is the sole surviving native mammal. The Jamaican rice rat and the long-tongued bat have become extinct in recent times as have half a dozen species of birds and several species of snakes and lizards. On Haiti, the Dominican Republic, Puerto Rico, and other islands of the Caribbean similar annihilations have occurred while during the same period the islands have burgeoned with people. Only in the past few years has there been concern of the islanders to husband their fast-dwindling natural resources.

Before man inhabits them, islands are the safest of all the habitats in the world. But as soon as man arrives, natural balances are destroyed, and living becomes hazardous. On islands large and small throughout the world, man's mark of havoc has been tragically shameful. Man destroys not only by altering the environment but also by proving himself to be the most successful and also the most sinister of all predators on earth.

THE NORTH WOODS

BALD EAGLE

MOOSE–
North American coniferous forest

The largest forest in the world circles the earth in the Northern Hemisphere in a broad belt commencing just south of the tundra. It is composed almost wholly of needle-leaved trees—pines, hemlocks, firs, spruces, and other conifers. Along the northern fringe, the trees straggle into the tundra, dwarfed and wind-warped in their struggle to grow in the cold land. Everywhere the forest has a sameness in appearance because the trees are all much alike. Where the trees are dense, a thick carpet of needles has built up beneath them over the centuries, and a gray soil forms slowly under this cover.

In Europe, this forest is called the taiga, occupying a vast lowland that was leveled by the great glaciers of the Ice Age and gouged deeply in places to form bogs, swamps, ponds, and lakes. In North America, the forest covers most of Canada. It fingers into the hardwood deciduous forest region of central United States and plunges even farther south along the mountain ridges.

Winters are long in the North Woods, lasting from six months in the south to as long as nine months along the tundra. Plants must grow and produce their seeds rapidly in the short summers. Animals, too, are hurried to find food to tide them over the winter months when food is scarce. Some animals migrate from the region in winter; others hibernate. Insects enter a dormant stage of their life history. But many kinds of animals remain active all winter long. As in summer, they feed on the seeds in the pine cones, or they eat bark or needles. And every plant eater is a potential meal for some predator.

Coniferous forests are the world's greatest source of timber. The harvesting has not always been done wisely. Too often there has been total destruction of the habitat over a wide area, unnecessarily disrupting or destroying wildlife populations. Even so, this great forest is one of the least disturbed of the large habitats, especially in North America, Many animals that once ranged widely over temperate regions have turned to the North Woods as a place of refuge. This is, for example, the last stronghold for the timber wolf. The wolf is not basically a woodland animal, but it escaped to the dense forests when the pressures of civilization became too great for it in its more preferred open lands to the south.

The lynx is another of the predators that was forced into retreat. In Europe, where the lynx once roamed widely, this wily cat has been hunted since ancient times. It is nearly extinct, found only in the most remote areas, but it is truly astonishing that it has survived at all. The lynx is more common in the North Woods of North America where the wilderness is much more extensive.

The wolverine, too, once ranged over a wide territory, though it has never been numerous. A terror even to other animals when it is competing for the same food, the wolverine has only one real enemy: man. Wolverines are killed for their fur, which is thick and warm, and its unusual ability to resist forming frost makes it a favorite for parkas. The wolverine is killed, too, because it is an aggressive competitor of man in the wild. Moving boldly just ahead of a trapper, a wolverine will rob traps of their catches and then hurry back to the trapper's camp and ransack it destructively. Relatives of the wolverine in the North Woods are sables, martens, mink, fishers, and weasels—all of them active carnivores and also sought by man for their fur. When the short-tailed weasel turns white in winter, it becomes the coveted ermine. Despite heavy trapping and hunting, none of these animals is endangered at present. But fads in furs or foolish "vermin extermination" programs could bring about swift elimination.

Bears, too, once roamed over most of the temperate regions of the Northern Hemisphere. Most of them are limited now to the North Woods. Bears are carnivores, but almost all of them round out their diets with fruits, seeds, berries, leaves, and even roots. Some kinds also eat fish. In winter, they sleep for long periods, but they do not hibernate—that is, there is no major change in their internal body activity. They may rouse several times during the winter to eat and then go back to sleep again.

In North America, black bears are the most common, living not only in the North Woods but also ranging much farther south—even into the Everglades. Only a bluish-coated subspecies inhabitating the mountains of Alaska is threatened. The grizzly bear, which may be eight feet tall when standing on its hind feet and may weigh as much as a thousand pounds, has become threatened with extinction. Since the settlement of the continent, the grizzly has been a prime quarry, killed for its meat and its hide and also to exterminate it as a beast too large and too dangerous to have around settlements. Only a hundred or so remain of a population that was once calculated to be in the thousands. They are protected only in national parks and reserves, however. In many other areas, they are still hunted for sport and also under the pretense that they have become a threat to livestock. Farther north, the slightly larger and closely related brown bear still persists but will be hard pressed for survival as the area is settled. A smaller brown bear lives in northern Europe and in Asia.

Beaver were once extremely common in most of the temperate regions of the Northern Hemisphere. Their hides were the "money" used in the frontier, and as a result, beaver were hunted and trapped to near extinction. Beaver are now abundant only in northern wilderness regions, especially in Europe and Asia.

Beaver feed on the bark and tender young twigs and branches of aspen, birches, and similar trees growing along the edges of waterways. With their sharp, chisel-like teeth, they can cut down sizable trees. They strip them of their branches and cut the trunks into smaller, more manageable pieces. The logs and larger branches are used in making their dams, the wood piled together tightly and then plastered heavily with mud. Digging to get the mud makes a pool on the upstream side of the dam. Smaller branches are stored in the deep pool behind the dam as a supply of food for use in winter. In the pools, too, the beaver build their lodge. Typically it has several underwater entrances to its single above-water living area, which has brush-covered vents at the top. After a period of years, the pool fills with silt, and the beaver then move off to make a new dam and pool. Behind them they leave a rich, moist meadow, a new habitat for wildlife.

The beaver is apparently now safe from extermination and is even returning to some areas from which it has been gone for many years. It is now also protected by law. But we must remember that the beaver was becoming scarce and might well have joined the passenger pigeon, the heath hen, and others if fashion had not changed and made its pelt no longer popular. The beaver was not spared because of a concern for the animal.

Birds of prey—hawks, eagles, owls—also haunt the coniferous forests, but where the trees are dense, it is difficult for the larger birds to hunt on the wing. Birds with broad wings live mainly along the coasts, at the edges of the forests, and in the few open areas. Like other animals, some of the birds—such as the giant eagles—have moved into the forests to escape persecution in the more open, more settled regions to the south.

Great Slave Lake in northern Canada is the nesting ground for the few remaining whooping cranes. Again, this is a much-repeated tale of slaughter taking a species to the brink of extinction. Year after year these magnificent birds, which stand five feet tall and have a wingspread of seven feet, made a 2,500-mile migration flight—from the Canadian wilderness to Mexico and the Gulf coast of Texas in autumn and then back again to their nesting ground in spring. All along the way they were met by guns, the killing for no better reason than to see the big beautiful birds tumble from the sky.

In days long gone, whooping cranes were apparently fairly common from Florida to California, and they probably nested in the northern part of the Great Plains. But they were pushed aside quickly by civilization. In recent times, their population may never have been greater than a few thousand birds, and it did not take the killing of many to reduce the number to a critical low. Real concern did not come, however, until only about 20 birds were left.

Protection finally began in the late 1930's, and few people believed that the species could be saved. A refuge was established for them at Aransas, Texas, where they wintered, and then conservationists set out to find where these few

BEAVER

birds nested. Happily, the site was located in what is now Canada's gigantic Wood Buffalo National Park. The major problem then became one of protecting the birds on their migration flights. Not everyone is sympathetic or understanding when it comes to preventing the extinction of a species.

Also tried—with some success—was a program of hatching eggs and rearing the chicks in captivity for release in the wild. This was believed worthwhile because even one natural disaster, such as a severe storm or a suddenly rampaging disease, could easily eliminate the few birds in the wild. The hatching program, conducted at the Patuxent Wildlife Research Center in Maryland, is an assurance program that the species will be saved—or hopefully so. Over the past several years, the number of whooping cranes has increased one year and then slipped back the next. There has been no steady upward climb yet. But today there are about 50 birds, and hopes are high that the whooping crane has been rescued from extinction.

The Siberian white crane, a counterpart of the whooping crane of North America, has also been so greatly reduced in numbers that it is now endangered. Fewer than two hundred of these big birds now exist, and like the whooping crane, they must run a gamut of guns and other hazards twice a year on their migration flights from the northern coniferous forest regions to the continent's warm southern coastal areas where they winter.

Kirtland's warbler, one of the rarest of all birds in North America, owes its plight only indirectly to man. Kirtland's warbler spends its summers in the coniferous forests of northern Michigan and winters in the Bahamas. As the forests were cleared, the common cowbird became more abundant in the same areas where the warbler lived, and the female cowbirds began centering their attention on the warbler's nest. Female cowbirds would lay a single egg in the nest of a Kirtland's warbler and then depart, leaving the little warbler with the chore of incubating the egg and then rearing the young cowbird. The warbler's own young, if any of the eggs ever hatched, were crowded out of the nest or starved because the aggressive little cowbird took all of the food. The result was a healthy, well-fed young cowbird but no warblers. This was the major factor contributing to the near-extinction of Kirtland's warbler.

In a dense forest, such as the warbler enjoyed originally, the cowbird would be less bothersome. For this reason, the U.S. Forest Service has set aside 4,000 acres of forest land where Kirtland's warbler nests in northern Michigan. It is hoped that the little bird's population will again increase.

KIRTLAND'S WARBLER

THE ENDS OF THE EARTH

EMPEROR PENGUIN—Antarctica

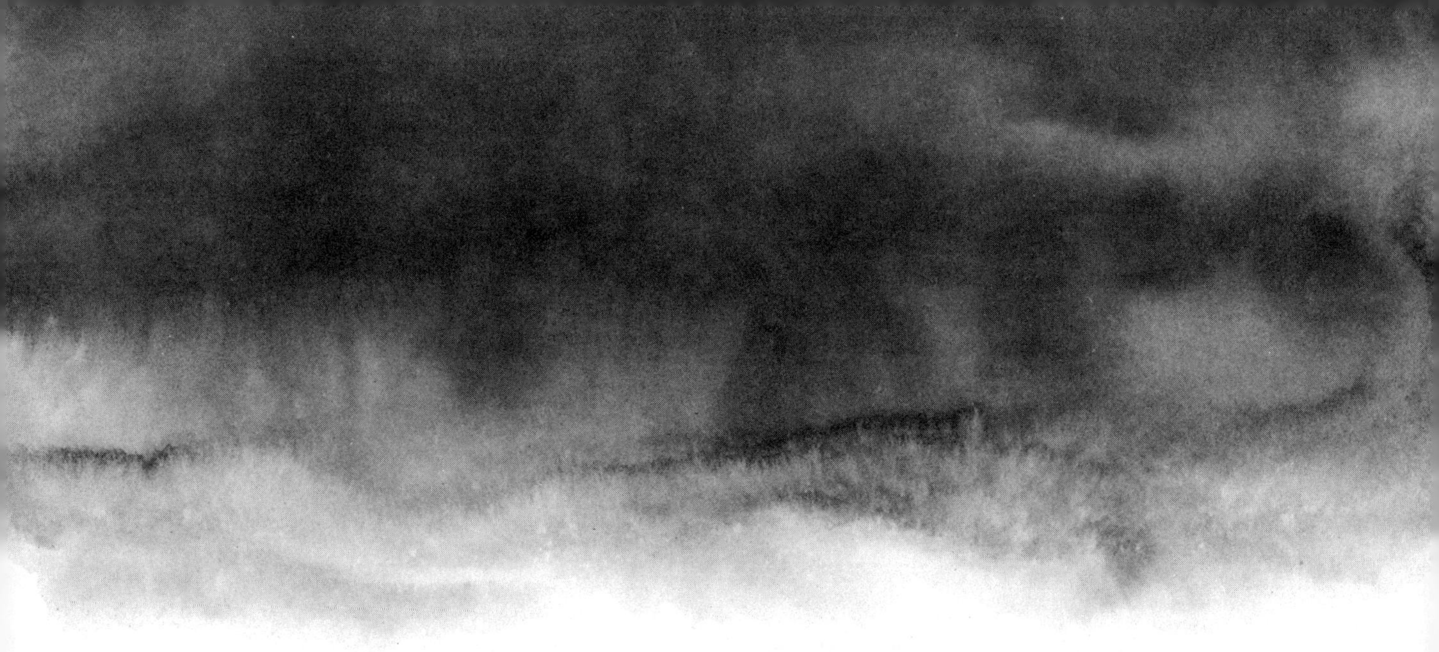

The Last Continent

ANTARCTICA, the last continent to be discovered and explored, is a vast land of ice and snow surrounding the South Pole. It is separated from all other land masses by the frigid southern extensions of the Indian, Pacific, and Atlantic oceans, so distinctive in character that they are commonly referred to as the Antarctic Ocean. Until about 1820, this 5½ million square miles was untouched by man. Its stormy, icy seas were too dangerous even to attract the insatiable probing of man. But it was exploitation that finally led man even to this desolate part of the world, for the waters surrounding the continent were first braved by seal and whale hunters.

Roughly circular in shape. Antarctica has two deep inlets—the Weddell Sea and the Ross Sea. The slim arm of the Palmer Peninsula stretches to within 650 miles of the tip of South America. In places on the continent, which may actually consist of many islands connected by ice, the covering of ice may be nearly two miles thick. Some of the mountain peaks are as much as 20,000 feet above sea level, among the tallest in the world.

Winter temperatures at the South Pole, located at about the center of the continent, may be 100 degrees below zero, and the sea freezes over for hundreds of miles from the coast. In summer, the temperature rises above freezing in many areas but is still much lower than in the Arctic. Along the coasts, great chunks of ice break off and float out to sea. The Antarctic icebergs are low and flat-topped in contrast to the land-formed, jagged-peaked icebergs of the Arctic. Sometimes these Antarctic icebergs drift as far north as Uruguay along the coast of South America before the warmer waters melt them.

The interior of Antarctica is a virtual desert—cold, stormy, wind-swept, and uninhabitable. Man is the only animal that has ever probed the depths of the great continent. Life is confined to the surrounding sea, the coasts, and the nearby islands. The sea is astonishingly rich with life because of a convergence of warm currents near the continent. It is surprising that several species of mosses and lichens and two kinds of flowering plants do exist along the shore. Microscopic plants or plankton are profusely abundant in the offshore seas in summer.

A few kinds of insects also inhabit the land, but only a few days every year are warm enough for them to be active. Rotifers, springtails, mites, and protozoans are found along the coasts, but they spend months or even years in a frozen state. The most active invertebrates are the parasites that live on some of the sea mammals and birds. Their micro-world is as warm as the body of their hosts.

Emperor and Adélie penguins and the skua gull are permanent residents along the coasts and on nearby islands. Several other kinds of penguins plus gulls, petrels, albatrosses, shearwaters, and snow buntings appear during the short summers. The elephant seal, leopard seal, Ross' seal, crabeater seal, humpback whale, finback whale, killer whale, sperm whale, blue whale, and several kinds of dolphins enter Antarctic waters when the ice breaks up in summer. Many of these animals lived here in far greater numbers in days gone by, but even in this most remote part of the world, man's exploiting invasions have had effect.

Only the hostility of the Antarctic has spared these animals from annihilation. Even so, the blue whale is now an endangered species. In winter, this great beast—the largest animal that has ever lived—moves northward into warmer waters, but it takes little or no food on these northward treks. It feeds when it returns to the Antarctic where it fattens on the abundance of krill, a shrimplike

ADÉLIE PENGUIN

planktonic crustacean. Since about 1900, hunters have reduced the world population of blue whales from about 500,000 to an estimated 1,000. Whether these animals can survive no matter what protection is given them is now in doubt, for the blue whale population may have reached that critical low point from which it cannot recover. And even if the whaling is stopped, the continued pollution of the seas poses a threat to their food supply.

Colonies of penguins generally consist of many thousands of birds. Of about 18 species that range into the Antarctic, only five kinds actually breed on the continent. Fortunately, man has not yet found an exploitable use for these unusual, flightless birds. Short of an indirect upset of their food supply, their chances for survival are good.

After the female emperor penguin lays her single egg, sometime during the month of May which is the beginning of the Antarctic winter, the male takes over the task of incubating it. Many males huddle together in the winter darkness while performing this chore, a two-month, foodless stint. Each male stands sluggishly and stubbornly over his egg, which is wrapped in a warm, blanket-like fold of skin between his feet. Meanwhile, the females are out at sea, stuffing their stomachs on squid. They come back just in time to claim their offspring and to take over the feeding of the newly hatched chicks. The males, relieved of their responsibility, then go out to sea themselves to regain the weight lost during the long period of fasting. The greatest threat forseeable for these birds is that their food may become contaminated with insecticides, as has happened with many other birds that live on food from the sea.

How long can the Antarctic continue to be a haven for whales, dolphins, seals, penguins, and other animals? The wide-ranging whales and dolphins have already suffered because they contact man in other seas. And even within their Antarctic realm, the animals are affected by the global pollution of the seas. Seals that have managed so far to escape mass massacre may yet fall victim to the hunters if world political squabbles are not resolved to preserve these valuable living resources.

A future certainty seems to be that man will become increasingly active in the Antarctic, conducting great numbers of scientific investigations, searching for natural resources, and sending a larger number of flights over the Antarctic wastelands on their shortest routes from Australia and New Zealand to South America. Already there are regular tourist excursions to Antarctica. Engineers and geologists have also appraised the continent's wealth in resources. Oil and natural gas will probably be the first of the resources to be exploited, and there are also valuable minerals. These developments must be watched carefully, for it seems up to now that wherever man has begun to dominate and control the environment, other forms of life have suffered. Life is sparse in this brutally cold region. Those animals that have managed to adapt to these severe conditions for survival should be spared. Antarctica is the only continent where no animal species has to date become extinct as a result of man. One can only hope the record will be maintained.

The Top of the World

The Arctic is not as well defined a region as the Antarctic, for there is no single, distinctive land mass. Rather, the North Pole consists of a sea of ice—in some places no more than a few feet thick but in others to a hundred or more feet deep. On land areas, glaciers form, some becoming thousands of feet in depth. They slide out into the sea as the peaked icebergs that are characteristic of the Arctic. In summer, the edges of the polar ice raft may break off also, but during the winter, the ice raft over the North Pole freezes solid again. The lands that lie closest to the Arctic Circle include parts of North America, Europe, and Asia, plus numerous islands. The largest of the islands are Ellesmere Island and Greenland, a tip of which is only about 450 miles from the North Pole. This is the northernmost piece of land.

Life on land is much richer in the milder, more calm Arctic region than in the Antarctic. The land surrounding the North Pole is the vast treeless tundra, carpeted with snow for six to nine months of the year. Since before recorded history, Eskimos have inhabited the Arctic region as have also the Lapps of northern Europe. Arctic seas, however, are constantly cold, lacking the benefit of warm currents that stir in nutrients and enrich the Antarctic. Life is less abundant in these waters.

RED-BREASTED MERGANSERS

Tundra lands are permanently frozen deep down—to almost half a mile in some places—but in summer, the top few inches thaw. Then the tundra burgeons with life. Literally thousands of species of lichens, mosses, and flowering plants grow in the boggy soil, kept moist because the frozen earth below prevents deeper percolation. Rapidly, the seemingly barren land first turns green and then becomes multicolored with blooming flowers while the sun shines the clock around at the peak of summer. The growing season is brief, and no time is lost in the production of seeds or other fruiting bodies. During this period the tundra becomes alive with animals. Insects appear in plaguing abundance. Even two species of frogs and a lizzard range into the Arctic.

Warm-blooded animals are also abundant in the tundra lands. Some move in from the south in summer and leave again before winter sets in. Among these familiar migrants are the caribou that travel more than 700 miles annually. Many birds nest in the tundra but spend the winters farther south. Other animals are year-round residents, specially adapted for survival in the severe winter.

Mammals wear extra-thick coats of fur, and their feet are heavily padded. Birds are thickly feathered, even on their feet. Both the fur and the feathers trap insulating pockets of air, preventing the rapid escape of heat radiated from the body. The ears of animals living in the Arctic are smaller than are those of their nearest relatives of temperate regions. Whales, seals, and walruses—the marine mammals of polar seas—are wrapped internally in a blanket of blubber.

Exposure to the cold is natural for these animals. They are equipped to cope with the weather—but not with man. Habitat destruction plays only a minor role in the Arctic regions. Hunting and trapping have been most damaging to these animals.

Polar bears, their total population down to perhaps as few as 10,000 animals, are killed almost wholly for sport—to see who can get the largest or to have some portion of the animal for a trophy. In times past, polar bear hunting did indeed require endurance and hardship. First came a long voyage by ship through the Arctic waters and then a trip by dog sled into the haunts of the polar bear. There the hunter met the powerful half-ton beast on its own ground, knowing full well that the animal could shred him and all of his trappings unless felled by the first bullet. In these contests, some hunters lost.

But today's hunter not only has a much more powerful, rapid-firing weapon but also does most of his hunting from an airplane. When a bear is sighted, the pilot lands and lets the hunter position himself—usually under a camouflage of white so that the bear will not even suspect his presence. Then the pilot takes off again and flies low over the big bear to herd it to the hunter and its doom. As many as a thousand bears are killed annually, most of them much in this manner. Protective laws are ignored.

Already living in a greatly reduced range, having once roamed far south into Iceland and even into Manchuria in Asia, the polar bears' existence is now threatened. If the hunters do not get them, then it seems that encroaching civilization will.

Steller's sea cow is an example of how quickly a species can be annihilated. These giant animals, to 20 feet long and weighing as much as four tons, were more than twice the size of the sea cows or manatees of the tropics and subtropics. They were unknown until 1741 when a Russian expedition found them in the waters around some islands in the Bering Strait. They were named for George Steller, the German naturalist serving as chief scientist on the expedition. Forced to spend the winter on the islands after their ship wrecked, the explorers tried eating all sorts of mammals, birds, and fishes before finally turning to the most delicately flavored of all—the huge, ugly, wrinkly sea cows, as tasty as their near relatives of warm seas.

The many explorers, whalers, and seal hunters who followed that first expedition into the region all took their share of the sea cows. By 1770, Steller's sea cow had disappeared. The total original population was never large —probably no more than 2,000 animals. Always there was hope that at least a few might have escaped—at least enough to build the population again. This hope was bolstered by occasional reports that some of the animals might have survived, some of the stories of sightings coming as late as the 1960's.

POLAR BEAR

Another victim of man in the same area was the spectacled cormorant, described in 1741 by the same scientist who also discovered Steller's sea cow. This big cormorant, which flew only poorly, lived on the Komandorskie Islands in the Bering Sea. This was another of the animals turned to as food by members of the expedition marooned on the islands after shipwreck. A ring of bare skin around each eye formed the "spectacles" that gave the birds their name. Unfortunately, the cormorants showed no fear of man and moved clumsily on land. They were easily killed with clubs, and one of the 15-pound birds made a meal for several men.

Starting with the explorers, the spectacled cormorants were either shot or clubbed into oblivion by the various visitors to the islands. By the mid-1800's, they were all gone. Today they are known only from a few specimens in museums.

All of the Arctic animals will suffer unless international agreements for their protection are arrived at and then enforced. From the giant musk oxen and caribou to wolves, foxes, and ermine—one by one they will be pushed to the brink and possibly into oblivion. Only a few hundred of the giant musk oxen are left now in North America, and there are none on the tundra lands of Asia and Europe. The largest number—perhaps as many as 20,000—are on Greenland and nearby islands. Now protected, these great shaggy beasts, weighing more than 600 pounds, are making a slow comeback. Some of the areas where they were exterminated are being repopulated. They are being captured and returned even to Asia, where they were originally native. The musk ox does not do well in captivity, and attempts to domesticate it have so far failed.

The caribou, in contrast, did yield to domestication. They are the reindeer of the Lapps in northern Europe. In the wild, the caribou are not faring as well. Their abundance in days gone by has been compared to the great herds of bison on the Great Plains, some saying their original population may have been as

TRUMPETER SWAN

great as 100 million. They supported the Eskimos and Indians, who followed them on their migration treks. But the European hunters soon began to whittle the herds to fractional size. In 1910, their population was estimated to be about 30 million. By 1940, this was reduced to 2 million; by 1955, to 275,000. The Canadian government set out to save the caribou in crisis, their studies indicating that the biggest factor contributing to the decline of the caribou was hunting by man. Laws were passed immediately to prevent future killing except by people living in remote areas who depended on the caribou for food. Enforcement has been difficult, but the wasteful shooting of animals for sport alone has apparently been stopped. Like the musk ox, the caribou is peculiarly adapted for life in the tundra. No other grazing animal except the musk ox can subsist on a diet of lichens and tough grasses. By proper management of its herds, the caribou can contribute significantly to man.

The many kinds of birds that move into the tundra will not be safe either if the region is disturbed. Many have already been reduced in numbers. The Eskimo curlew was one of the first to approach extinction. Its massacre did not take place in the Arctic, however. This small shorebird, closely related to sandpipers, once nested in the tundra by the countless thousands. In winter it migrated southward to the grasslands of Argentina. On its migration flight, it flew from Newfoundland and Labrador almost directly southward over the sea, some of the huge flocks following the coast. The return trip was made mostly overland—through South America, then across the Gulf of Mexico and up the Mississippi River Valley into the tundra land.

By 1900 the Eskimo curlew had become scarce. Wherever the migrating flocks came close to shore on their southward flight and throughout their return trip up the central part of the continent, they were greeted by guns. Hordes of hunters downed the birds by the thousands and hauled them away to markets by the wagonload. Many that were shot were never retrieved. Others were collected only to be fed to pigs. No one was concerned because the birds appeared to exist in limitless numbers, but there was a limit, as there was also with the dodo, the passenger pigeon, and others. Some say the Eskimo curlew has long been annihilated. Others hold out hope that at least a few persist. Their nesting site in the tundra still awaits them, so far little altered.

The whistling swan nests in the tundra, and like other large waterfowl, it has been heavily hunted. Today it is rare. Hopefully, it may be rescued from extinction as was the even larger trumpeter swan that in 1900 was believed to be extinct. Then a few survivors were found in Yellowstone National Park. These birds were given complete protection, and now several thousand trumpeters exist in Yellowstone and in other parks in the United States and in Canada.

On a migration flight spanning nearly 25,000 miles annually, the Arctic tern journeys from one polar sea to the other, mostly over the open ocean and hence safe from man. But like all other sea birds, the Arctic tern depends on the seas for its food and will remain safe only as long the seas are not totally polluted. The signs now are not good.

A LAST CHANCE—WITH NEW HOPE

CAROLINA PARAKEET

WHOOPING CRANE

IVORY-BILLED WOODPECKER

ASIATIC LION

WE CANNOT bring the dodo back from the dead. It is gone forever, as are the great auk, quagga, passenger pigeon, Steller's sea cow, and more than 350 other victims of man's compulsion to demonstrate his dominance over nature and all the creatures that share the world with him. First, perhaps, man fought wild beasts primarily from fear. Then they became his food. After he mastered tilling the land and husbanding the dozen or so domesticated animals, he continued hunting for sport—even until today. And all the while man himself has proliferated, tearing down nature as he has built his own kind of world and forgetting that he is part of that natural world himself. Finally he has come to the realization that even his mightiest cities offer only thin-shelled separation from nature. He knows now that however remote he tries to make himself, he is still an element in the various and intricate webs of life, no less so than the hippopotamus, the alligator, the pelican, and every other living creature.

When these animals die, the webs become somehow frayed or weakened; in some places they may be broken completely. Any such disturbance reverberates throughout the living world. This is well enough understood now so that people no longer look at nature as an enemy to be conquered but rather as the hub around which all life fits. Man has proven his power. Of all living things, he has established himself as the most dangerous, not only to other creatures but also to himself. This recognized, it has come time now to mend the many rents, to admit the causes of the current predicaments and to stop the senseless pursuit of wildlife into oblivion. Where it becomes a question of man versus beast, man's choice is obvious, but such a drastic and final alternative can be avoided if man uses the wits and wisdom that have set him apart from all other living things.

Massacres

Overhunting must be stopped. Controlled harvests—with a gun and for sport, if some insist that killing comes in this category—may be allowed on some lands to prevent overpopulation and to maintain healthy populations of animals. But however it is done, these harvests must be carefully managed.

The most devastating offenses are the killings done to supply fashion and curio markets. These slaughters could be stopped most effectively if people stopped buying, thus eliminating the market. A bracelet made from giraffe hair, shoes of crocodile leather, skin lotion of turtle oil, a coat from a leopard skin—every such item puts another species closer to extinction. If people vented their feelings to store owners and to suppliers, these products would soon disappear from the shelves. In the United States, the importation of products made from endangered species is illegal, but smuggling continues.

Poisons

Man has ravaged the natural world without using guns. Starting with DDT in World War II, he has built an arsenal of chemical poisons with which to do battle against insect pests. The good that has been done is not questioned. These chemicals have eliminated diseases from some areas of the world and made them inhabitable for the first time in history. They have also increased crop yields. Encouraged by the successes, man has increased the dosages and also, in some instances, the potency of the chemicals. Many millions of tons of pesticides have been spread and sprayed over agricultural and forest lands of the United States in the past quarter of a century. Other developed and fast-growing countries have been equally liberal in the use of chemicals.

The realization that pesticides might be doing more harm than good came slowly, but it is now clearly documented. Many of the chemicals do not make their kill and then degenerate into some harmless product. Rather, they retain their original or a related structure and also their potency so that wherever they exist as residues in nature they continue to pose a threat to living things. The case against DDT is clear, and it is now banned in the United States, although still in use elsewhere.

Washed from the land by rains, most of the DDT eventually finds its way into streams, lakes, and coastal waters. In places where heavy spraying was done to control forest insect pests or to eradicate plaguing pests of crops, kills of fish also occur. The fish may not turn belly up until months later—usually in winter when they begin living on the fat stored in their bodies. This fat has become lethal because from each insect or other morsel of food the fish has acquired a small amount of DDT. These have added up into a killing dosage.

This process, called biological magnification, has put the brown pelican, bald eagle, and other birds of prey on the list of endangered species. Almost every fish or shellfish eaten by these birds contains some of the poison, and it accumulates in the birds' bodies, either killing them directly or at least preventing them from producing young. Most commonly the females lay thin-shelled eggs that never hatch, and so year after year, there is no increase in the population.

In food pyramids, the predators, such as the bald eagle and the peregrine falcon, form the peak. They are at the top of the heap, so to speak. But in the poisoning by pesticides, they are typically the first to be affected because they get magnified dosages of the poison. Meanwhile, the pest species that man set out to control commonly builds a population resistant to the poison and exists in more abundance than before, many of its natural enemies having been eliminated or much reduced in numbers.

One of the hopes for wildlife is that man has learned this lesson soon enough to stop using pesticides in such great abundance. Even so, it will take many decades for the effects of its use in the past to disappear. The threat to wildlife and eventually to man himself due to these poisons is certainly no longer debatable. These poisons, along with other pollutants, must be eliminated from the environment—and not only to spare wildlife but also to assure man himself of survival.

Destruction of Habitat

Habitats are made unlivable both by poisoning and by physical destruction, which is probably the greatest threat to wildlife today. As man has cleared the land and converted it to farming or living space for himself, many species of wildlife have been forced to retreat. But civilization's so-called advance continues. For many of the now endangered species, their last stronghold on earth is being battered down.

In some instances, the dilemma is not avoidable. If man truly needs the land to feed and house people, the wild inhabitants must move on to other places. But in many cases man has simply plundered the natural world.

Pillaging of this sort is going on now in the fragile tropics of South America. Vast areas are being leveled to grow crops. The technique is simple: an area is cut over and then burned. Crops are then planted, but the soil is neither very rich nor deep. Exposed to the heavy, leaching rains and the burning sun, the exposed land is quickly impoverished and is abandoned. Then another tract is laid bare to the elements, and parcel by parcel, the wildlife is destroyed as their total available living space becomes less and less.

In a world of many governments and diverse opinions, hungry people turn to the most expedient ways of feeding themselves. They have no concern about

A MODERN ZOO

the past or the future. No one can really expect these people to give thought to living space for colorful hummingbirds, a spotted cat, or any other wild creatures as long as their own stomachs are sore with hunger. Only well fed and satisfied people can become emotionally disturbed about the plight of wild animals. But the time is fast coming when world sharing of food and other resources will be necessary. Then it will be easier to communicate the need for concern about living spaces for all animals and plants on earth. It will be more easily understood that we all exist in really one world—the biosphere, or that part of the earth that is inhabitable.

The New Zoo

Conservationists are generally cheered by the philosophy of the new zoos appearing throughout the world. For many species of animals, these zoos represent a last chance on earth.

Zoos used to be—and some, regrettably, still are—nothing more than miserable prisons for wild animals. The captives were simply caged and put on exhibit where they could be poked at, screamed at, fed all sorts of trash, and allowed to wallow in filth and become diseased. Only the most hardy survived long in these conditions. Those that succumbed were quickly replaced by another of the same kind, a large number of the animals dying in shipment or in waiting reservoirs before ever reaching their destination.

Collecting animals for zoos was a profitable business, and zoos were directly responsible for the decline in numbers of species that were never abundant originally. The zoos took great pride in exhibiting specimens of the most rare animals and thus contributed to their becoming even more rare. Sharing this dubious distinction were pet dealers who offered exotic pets for sale—and still do. Nor were scientists less considerate of their fellow creatures on earth. Medical research has contributed greatly to the near-extinction of a number of species, particularly the primates. They are supplied now mainly from laboratories where the animals are bred specifically for this purpose, but conservationists remain on the alert at traditional shipping centers around the world.

New zoos are near-duplications of the animals' original habitat, providing all of their good features and excluding most of the hazards. Visitors are, in most places, no longer within reach of the animals and sometimes cannot even be seen by them, yet they see the animals eating, moving about, and resting in surroundings that are almost identical to what they would have in the wild. In the case of large and potentially dangerous animals, they are separated from them by walls too steep to be scaled and by deep moats, both designed to fit into the landscape. Similar kinds of barriers prevent predators and prey from intermingling. Only animals that are compatible are put in the same enclosure.

Here and there throughout the zoo are interpretation centers where various audio-visual aids, including TV cameras, the various exhibits and the behavior of the animals are explained. All of the maintenance and feeding are taken care of from areas that are concealed from the visitors so that the undisturbed natural atmosphere is maintained.

Nocturnal animals live in similar conditions that provide them with a natural habitat, but they are kept in windowless buildings with artificial lighting that reverses night and day for them. Thus, they are active during the day when they can be observed by visitors. Their habitats are lighted with low-intensity red lights that the animals themselves cannot see. At night, when the zoo is closed to visitors, the world of the nocturnal animals will blaze with bright lights to send them into their sleeping period.

In free-flight aviaries, visitors can walk through the exhibit at different levels, even at treetop height, to watch birds in a large enclosure in which they can fly freely. Tunnel-like approaches permit visitors to observe some animals at close range in their habitat from glass panels that allow vision in one direction only. Other tunnels lead into underwater observation "bells" for watching otters, beavers, waterfowl, or fish moving about in their below-the-surface abodes.

In all exhibits, in fact, supreme effort is expended toward creating conditions of comfort for the animals with a minimum of disturbance. And all the while, of course, the animals are being closely observed by the zoo scientists to make certain they are well fed and in good health. At the slightest suggestion of illness an animal is given medical attention. Health-wise, certainly, zoo animals are better off than those that live in the wild, and they live much longer lives.

Visitors to a zoo who are never satisfied unless their trip has included petting an animal enjoy the so-called "touch" areas where docile animals are kept, most of them within reach. Some of these animals are common domesticated types, for there are many city dwellers who have no other opportunity to see ducks, chickens, cows, pigs, horses, and similar longtime companions of man.

These various steps have put the new and modern zoos in the category of educational institutions for learning about animals and their behavior. Zoos are also laboratories for the serious students who want to learn more about the

NORTH AMERICAN FLYING SQUIRREL

habits, nutritional needs, and diseases of the various species. This is of great importance in knowing how to provide for the needs of animals in the wild. In all cases, the most sophisticated of modern equipment is being utilized to achieve a maximum benefit from these opportunities.

Most important for endangered species, zoos have become breeding grounds where the species are maintained. Sometimes sufficient surpluses are produced for release in the wild to help restore diminished numbers there. For many species, at least, the taking of more animals from the wild is no longer necessary. The zoos have learned how to supply their own needs.

Some zoos have also specialized. Rather than trying to exhibit a great variety of animals and doing so poorly, they concentrate on showing only particular species or groups of animals. In this way they can utilize a limited space more effectively and can also do an exceptional job of education. Some exhibit only animals that are native to their area or their country, the exhibits and the information provided to visitors adding greatly to the appreciation and understanding of these animals in the wild. Still others have specialized by showing only endangered species, while at the same time working closely with their governments to make their efforts a genuine contribution to the preservation of these species.

Zoos have unquestionably become centers for educating both the public and the scientist in the ways of wildlife. The importance of this role is destined to grow rather than to diminish as time passes.

AFRICAN BUSHBABY

National Parks and Sanctuaries

Every year more lands are set aside as national parks and sanctuaries for wildlife. World-wide there are now several hundred such areas, some of them covering thousands of square miles. Here people can see and appreciate wildlife in natural conditions while at the same time renewing their own kinship with the natural world. Population surpluses from some of these areas feed into the surrounding lands, and the core areas maintain a protected gene pool of animals and also plants that are native to the region. Here scientists can work in undisturbed environments to determine the wisest use of the land for all living things. Even for science, this understanding of the natural world is a continuing and ever-revealing process.

Obviously, there must be extensive portions of these restricted lands where no humans can trespass—that is, no sightseers or others who make their visits only for personal appreciation and enjoyment. These virtually untouched areas become the most valued elements of the large, living laboratories of the natural environment. They are places where scientists can observe the way nature works without the meddling of man. It is true, too, that some kinds of wildlife cannot tolerate intrusions.

In the early days, these lands for wildlife were established principally by the very rich or by royalty to supply game for hunting. They were like large ranches on which the livestock was wild animals. Now nearly all of the parks and preserves are government owned. Conservation of wildlife has indeed become a big and important business because of its close and clear relationship to the welfare of man.

Tourism, even of the gentle and kindly sort, has become one of the great threats to wildlife in this world of fast-traveling multitudes. Even wildest Africa is littered with discarded film cartons, flashbulbs, and other evidences of man's prying into the private lives of wild creatures. Unspoiled nature is indeed difficult to find anywhere. With four-wheel-drive vehicles, helicopters, lugged shoes, and backpacks—or whatever means are necessary—man invades virtually every out-of-the-way place on the globe.

Oddly, the effects of this kind of pollution on the natural world are paradoxical. On the one hand, it is distressingly destructive and disruptive, for the physical impact of the hordes of viewers, no matter how sincere their intentions, is both erosive and littering. Yet the exposure of the many people to the wonders of nature does cultivate much-needed understanding and appreciation of environmental problems. It is most obvious that tourism must be closely regulated to preserve wilderness areas for wildlife. Travel and advertising agencies must be keenly aware of the roles they play in environmental protection. Many are now aware of how they might contribute to ecological disasters and are exercising at least some degree of control over where they steer people to go on excursions.

Gene Banks

One of the most unusual efforts being made to preserve endangered species is the "banking" of frozen genetic materials. This is an effort of a California scientist working in cooperation with the University of California, University of Washington, and the National Institutes of Health. So far, genetic samples have been taken from about a thousand animals, including the California condor, a number of reptiles and birds unique to the Galápagos Islands, and from many animals that have died in captivity. Soviet scientists have also promised to contribute materials from the next woolly mammoth that is discovered in the Siberian tundra. All of these samples are frozen quickly and deeply, then divided into three portions for storage in three different locations as a precaution in case of power failures.

What good is this material? Right now it has limited value, though geneticists can obtain small samples from the "banks" for their research where the material will contribute significantly to a study. But if scientists do learn how to generate living cells—and this possibility is imminent—then these genetic materials could be used to re-create whatever animals they came from. So these "banks" of genetic samples from now-doomed animals may make it possible for these same animals to stage dramatic comebacks in the future.

The Future?

There is still hope for endangered animals. Their worlds have changed and will never again be the same, as the world has changed also for man. Man is peculiar among animals in being able to adapt to a great variety of conditions. If necessary, he can even surround himself with an artificial environment to continue his existence. Other animals are less flexible and thus more easily victimized by changes in the environment.

But this capacity that has taken man to the brink of disaster, along with about a thousand other species of animals, can also be turned in the opposite direction. Just as man, the thinker, has manipulated world environments into their present condition, verging on a global catastrophe, so he can alter the course and make the world livable again—for himself and for the wild creatures that share the world with him.

Endangered Wildlife

The animals named here appeared on either the U.S. or the international list of endangered species at the time this book was published. Some of the species listed may be extinct by the time you read the book; others may have been removed from the list, still others added. But every name is dramatic testimony to man's impact on his environment—and on the world of wildlife.

Not enumerated here are the growing numbers of invertebrates being added to the endangered category: at publication time, they included 24 species of clams native to North America, a tree snail found in the Admiralty Islands, and 6 species of butterflies from coastal California. Also excluded are more than 1,700 species of threatened plants.

UNITED STATES

FISHES

Bonytail, Pahranagat
(Gila robusta jordani)

Chub, humpback
(G. cypha)

Chub, Mohave
(Siphateles mohavensis)

Cisco, longjaw
(Coregonus alpenae)

Cui-ui *(Chasmistes cujus)*

Dace, Kendall Warm Springs
(Rhinichthys osculus thermalis)

Dace, Moapa
(Moapa coriacea)

Darter, bayou
(Etheostoma rubrum)

Darter, fountain
(Etheostoma fonticola)

Darter, Maryland
(E. sellare)

Darter, Okaloosa
(E. okaloosae)

Darter, snail
(Percina sp.)

Darter, watercress
(E. nuchale)

Gambusia, Big Bend
(Gambusia gaigei)

Gambusia, Clear Creek
(G. heterochir)

Gambusia, Pecos
(G. nobolis)

Killifish, Pahrump
(Empetrichthys latos)

Madtom, Scioto
(Noturus trautmani)

Pike, blue
(Stitzotedion vitreum glaucum)

Pupfish, Comanche Springs
(Cyprinodon elegans)

Pupfish, Devil's Hole
(C. diabolis)

Pupfish, Owen River
(C. radiosus)

Pupfish, Tecopa
(C. nevadensis calidae)

Pupfish, Warm Springs
(C. nevadensis pectoralis)

Squawfish, Colorado River
(Ptychocheilus lucius)

Stickleback, unarmored threespine
(Gasterosterus aculeatus williamsoni)

Sturgeon, shortnose
(Acipenser brevirostrum)

Topminnow, Gila
(Poeciliopsis occidentalis)

Trout, Arizona
(Salmo sp.)

Trout, Gila
(S. gilae)

Trout, greenback cutthroat
(S. clarki stomias)

Trout, Lahontan cutthroat
(S. clarki henshawi)

Trout, Paiute cutthroat
(S. clarki seleniris)

Woundfin
(Plagopherus argentissimus)

AMPHIBIANS AND REPTILES

Alligator, American
(Alligator mississipiensis)

Boa, Puerto Rican
(Epicrates inornatus)

Crocodile, American
(Crocodylus acutus)

Lizard, blunt-nosed leopard
(Crotaphytus silus)

Salamander, desert slender
(Batrachoseps aridus)

Salamander, Santa Cruz long-toed
(Ambystoma macrodactylum croceum)

Salamander, Texas blind
(Typhiomolge rathboni)

Snake, San Francisco garter
(Thamnophis sirtalis tetrataenia)

Toad, Houston
(Bufo houstonensis)

BIRDS

Akepa, Hawaii (akepa)
(Loxops coccinea)

Akepa, Maui (akepuie)
(L. coccinea ochraceú)

Akialoa, Kauai
(Hemignathus procerus)

Akiapolaau
(H. wilsoni)

Bobwhite, masked
(Colinus virginianus ridgwayi)

Condor, California
(Gymnogyps californianus)

Coot, Hawaiian
(Fulica americana alai)

Crane, Mississippi sandhill
(Grus canadensis pulla)

Crane, whooping
(G. americana)

Creeper, Hawaiian
(Loxops maculata mana)

Creeper, Molakai (kakawahie)
(Loxops maculata flammea)

Creeper, Oahu (alauwahio)
(L. maculata maculata)

Crow, Hawaiian (alala)
(Corvus tropicus)

Curlew, Eskimo
(Numenius borealis)

Duck, Hawaiian (koloa)
(Anas wyvilliana)

Duck, Laysan
(A. laysanensis)

Duck, Mexican
(A. diazi)

Eagle, southern bald
(Haliaeetus leucocephalus)

Falcon, American peregrine
(Falco peregrinus anatum)

Falcon, Arctic peregrine
(F. peregrinus tundrius)

Finches, Laysan and Nihoa
(Psittirostra cantans)

Gallinule, Hawaiian
(Gallinula chloropus sanvicensis)

Goose, Aleutian Canada
(Branta canadensis leucopareia)

Goose, Hawaiian (nene)
(B. sandvicensis)

Hawk, Hawaiian (io)
(Buteo solitarius)

Honeycreeper, crested (akohekohe)
(Palmeria dolei)

Kite, Florida Everglade (snail kite)
(Rostrhamus sociabilis plumbeus)

Millerbird, Nihoa
(Acrocephalus kingi)

Nukupuus, Kauai and Maui
(Hemignathus lucidus)

Oo, Kauai (oo aa)
(Moho braccatus)

Ou *(Psittirosta psittacea)*

Palila *(P. bailleui)*

Parrot, Puerto Rican
(Amazona vittata)

Parrotbill, Maui
(Pseudonestor xanthorphrys)

Pelican, brown
(Pelecanus occidentalis)

Petrel, Hawaiian dark-rumped
(Pterodroma phaeopygia sandwichensis)

Pigeon, Puerto Rican plain
(Columba inornata wetmorei)

Po'o uli
(Melamprosops phaeosoma)

Prairie chicken, Attwater's greater
(Tympanuchus cupido attwateri)

Rail, California clapper
(Rallus longirostris obsoletus)

Rail, light-footed clapper
(R. longirostris levipes)

Rail, Yuma clapper
(R. longirostris yumanensis)

Shearwater, Newell's manx
(Puffinus puffinus newelli)

Sparrow, Cape Sable
(Ammospiza mirabilis)

Sparrow, dusky seaside
(A. migrescens)

Sparrow, Santa Barbara
(Melospiza melodia graminea)

Stilt, Hawaiian
(Himantopus himantopus knudseni)

Tern, California least
(Sterna albifrons browni)

Thrush, large Kauai
(Phaeornis obscurus myadestina)

Thrush, Molokai (olomau)
(P. obscurus rutha)

Thrush, small Kauai (puaiohi)
(P. palmeri)

Warbler, Bachman's
(Vermivora bachmanii)

Warbler, Kirtland's
(Dendroica kirtlandii)

Whip-poor-will, Puerto Rican
(Caprimulgus noctitherus)

Woodpecker, ivory-billed
(Campephilus principalis)

Woodpecker, red-cockaded
(Dendrocopus borealis)

MAMMALS

Antelope, peninsular pronghorn
(Antilocapra americana peninsularis)

Bat, gray *(Myotis grisescens)*

Bat, Hawaiian hoary
(Lasiurus cinereus semotus)

Bat, Indiana
(Lyotis sodalis)

Bear, grizzly
(Ursus arctos horribilis)

Cougar, eastern
(Felis concolor cougar)

Deer, Cedros Island mule
(Odocoileus hemionus cerrosensis)

Deer, Columbian white-tailed
(Odocoileus virginianus leucurus)

Deer, Key
(O. virginianus clavium)

Ferret, black-footed
(Mustela nigripes)

Fox, San Joaquin kit
(Vulpes macrotis mutica)

Manatee, Florida (sea cow)
(Trichechus manatus latirostris)

Mouse, salt marsh harvest
(Reithrodontomus raviventris)

Panther, Florida
(Felis concolor coryi)

Prairie dog, Utah
(Cynomys parvidens)

Pronghorn, Sonoran
(Antilocapra americana sonoriensis)

Rat, Morro Bay kangaroo
(Dipodomys heermanni morroenis)

Squirrel, Delmarva Peninsula fox
(Sciurus niger cinereus)

Wolf, Eastern timber
(Canis lupus lycaon)

Wolf, Mexican *(C. lupus baileyi)*

Wolf, Northern Rocky Mountain
(C. lupus irremotus)

Wolf, red *(C. rufus)*

WORLD

(United States included only when species is shared with another country)

FISH

NAME	*WHERE FOUND*
Ala Balik *(Salmo platycephalus)*	Turkey.
Ayumodoki *(Hymenophysa curta)*	Japan.
Blindcat, Mexican *(Prietella phreatophila)*	Mexico.
Catfish *(Pangasius sanitwongsei)*	Thailand.
Catfish, Giant *(Pangasianodon gigas)*	Thailand.
Cicek *(Acanthorutilus handlirschi)*	Turkey.
Nekogigi *(Coreobagrus ichikawai)*	Japan.
Tanago, Miyako *(Tanakia tanago)*	Japan.

AMPHIBIANS

NAME	*WHERE FOUND*
Frog, Israel painted *(Discoglossus nigriventer)*	Israel.
Frog, Stephen Island *(Leiopelma hamiltoni)*	New Zealand.

REPTILES

NAME	*WHERE FOUND*
Boa, Jamaica *(Epicrates subflavus)*	Jamaica.
Caiman (see Yacare)	
Crocodile, Cuban *(Crocodylus rhombifer)*	Cuba.
Crocodile, Morelet's *(Crocodylus moreletii)*	Mexico, British Honduras, Guatemala.
Crocodile, Nile *(Crocodylus niloticus)*	Africa.
Crocodile, Orinoco *(Crocodylus intermedius)*	Orinoco River (northern South America).
Gavial *(Gavialis gangeticus)*	Pakistan, India, Burma, Bangladesh.
Gecko, Day *(Phelsuma newtoni)*	Mauritius.
Gecko, Round Island, day *(Phelsuma guentheri)*	Mauritius.
Iguana, Anegada ground *(Cyclura pinguis)*	Virgin Islands: Anegada Island.
Lizard, Barrington land *(Conolophus pallidus)*	Ecuador: Galápagos Islands.
Terrapin, River (Tuntong) *(Batagur baska)*	Burma, India, Indonesia, Malaysia, Bangladesh.

REPTILES (cont'd.)

NAME	*WHERE FOUND*
Tortoise, Galápagos *(Testudo elephantopus)*	Ecuador: Galápagos Islands.
Tortoise, Madagascar radiated *(Testudo radiata)*	Madagascar.
Tortoise, Short-necked or swamp *(Pseudemydura umbrina)*	Australia.
Tuatara *(Sphenodon punctatus)*	New Zealand.
Turtle, Aquatic box *(Terrapene coahuila)*	Mexico.
Turtle, Atlantic Ridley *(Lepidochelys kempii)*	Mexico.
Turtle, Hawksbill *(Eretmochelys imbricata)*	Tropical seas.
Turtle, Leatherback *(Dermochelys coriacea)*	Tropical seas and temperate seas.
Turtle, South American River *(Podocnemis expansa)*	Orinoco and Amazon River Basin (northern South America).
Turtle, South American River *(Podocnemis unifilis)*	Orinoco and Amazon River Basin (northern South America).
Yacare (Caiman) *(Caiman yacare)*	Bolivia, Argentina, Peru, Brazil.

BIRDS

NAME	*WHERE FOUND*
Albatross, Short-tailed *(Diomedea albatrus)*	Japan.
Bobwhite, Masked *(Colinus virginianus ridgwayi)*	United States, Mexico.
Bristlebird, Western *(Dasyornis brachypterus longirostris)*	Australia.
Bulbul, Mauritius olivaceous *(Hypsipetes borbonicus olivaceus)*	Mauritius.
Bullfinch, Sao Miguel *(Pyrrhula pyrrhula murina)*	Azores.
Bustard, Great Indian *(Choriotis nigriceps)*	India, Pakistan.
Cahow *(Pterodroma cahow)*	Bermuda.
Condor, Andean *(Vultur gryphus)*	Colombia to Chile, Argentina.
Crane, Hooded *(Grus monachus)*	Japan, U.S.S.R.
Crane, Japanese *(Grus japonensis)*	Japan, China, Korea, U.S.S.R.
Crane, Siberian white *(Grus leucogeranus)*	Siberia to India.
Crane, Whooping *(Grus americana)*	Canada, United States.
Cuckoo-shrike, Mauritius *(Coquus tropicus)*	Mauritius Island
Cuckoo-shrike, Reunion *(Coquus newtoni)*	Reunion Island

BIRDS (cont'd.)

NAME	WHERE FOUND
Curassow, Red-billed *(Crax blumenbachii)*	Brazil.
Curassow, Trinidad white-headed *(Pipile pipile pipile)*	Trinidad.
Curlew, Eskimo *(Numenius borealis)*	Canada to Argentina.
Dove, Cloven-feathered *(Drepanoptila holosericea)*	New Caledonia.
Dove, Grenada *(Leptotila wellsi)*	West Indies: Grenada.
Dove, Palau ground *(Gallicolumba canifrons)*	Palau Islands.
Duck, White-winged wood *(Cairina scutulata)*	India, Thailand, Malaysia, Burma, Indonesia (including Java).
Eagle, Monkey-eating *(Pithecophaga jefferyi)*	Philippines.
Eagle, Spanish imperial *(Aquila heliaca adalberti)*	Spain, Morocco, Algeria.
Egret, Chinese *(Egretta eulophotes)*	China, Korea.
Falcon, American peregrine *(Falco peregrinus anatum)*	Canada, United States, Mexico.
Falcon, Arctic *(Falco peregrinus tundrius)*	Canada, United States, Mexico.
Fantail, Palau *(Phipidura lepida)*	Palau Islands.
Flycatcher, Euler's *(Empidonax euleri johnstonei)*	West Indies: Grenada.
Flycatcher, Palau fantail *(Ripidura lepida)*	Palau Islands
Flycatcher, Seychelles black *(Terpsiphone corvina)*	Seychelles.
Flycatcher, Tahiti *(Pomarea nigra nigra)*	Tahiti.
Fody, Seychelles *(Foudia sechellarum)*	Seychelles.
Goose, Aleutian Canada *(Branta canadensis leucopareia)*	Japan, United States.
Goshawk, Christmas Island *(Accipiter fasciatus natalis)*	Indian Ocean: Christmas Island.
Grackle, Slender-billed *(Cassidix palustris)*	Mexico.
Grass-wren, Eyrean *(Amytornis goyderi)*	Australia.
Grebe, Atitlán *(Podilymbus gigas)*	Guatemala.
Guan, Horned *(Oreophasis derbianus)*	Guatemala, Mexico.
Gull, Audouin's *(Larus audouinii)*	Mediterranean.
Hawk, Anjouan Island sparrow *(Accipiter francesii pusillus)*	Comoro Islands.
Hawk, Galápagos *(Buteo galapagoensis)*	Ecuador: Galápagos Islands.

BIRDS (cont'd.)

NAME	WHERE FOUND
Honeyeater, Helmeted *(Meliphaga cassidix)*	Australia.
Ibis, Japanese crested *(Nipponia nippon)*	Japan, Korea, U.S.S.R., China.
Kagu *(Rhynochetos jubatus)*	New Caledonia.
Kakapo *(Strigops habroptilus)*	New Zealand.
Kestrel, Mauritius *(Falco punctatus)*	Mascarene Islands: Mauritius.
Kestrel, Seychelles *(Falco araea)*	Seychelles.
Kite, Cuba hook-billed *(Chondrohierax wilsonii)*	Cuba.
Kite, Grenada hook-billed *(Chondrohierax uncinatus mirus)*	West Indies: Grenada.
Kokako *(Callaeas cinerea)*	New Zealand.
Magpie-robin, Seychelles *(Copsychus seychellarum)*	Seychelles.
Malkoha, Red-faced *(Phaenicophaeus pyrrhocephalus)*	Ceylon.
Megapode, LaPerouse's *(Megapodius laperouse)*	Palau Islands, Mariana Islands.
Monarch, Tinian *(Monarcha takatsukasae)*	Mariana Islands: Tinian Island.
Ostrich, Arabian *(Struthio camelus syriacus)*	Jordan, Saudi Arabia.
Ostrich, West African *(Struthio camelus spatzi)*	Western Sahara.
Owl, Anjouan scops *(Otus rutilus capnodes)*	Comoro Islands.
Owl, Palau *(Otus podargina)*	Palau Island.
Owl, Seychelles *(Otus insularis)*	Seychelles.
Owlet, Mrs. Morden's *(Otus ireneae)*	Kenya.
Parakeet, Forbes' *(Cyanoramphus auriceps forbesi)*	New Zealand.
Parakeet, Golden-shouldered *(Psephotus chrysopterygius)*	Australia.
Parakeet, Mauritius ring-necked *(Psitiacula krameri echo)*	Mauritius.
Parakeet, Ochre-marked *(Pyrrhura cruentata)*	Brazil.
Parakeet, Paradise *(Psephotus pulcherrimus)*	Australia.
Parakeet, Scarlet-chested *(Neophema splendida)*	Australia
Parakeet, Turquoise *(Neophema pulchella)*	Australia.
Parrot, Bahamas *(Amazona leucocephala bahamensis)*	Bahamas.
Parrot, Ground *(Pezoporus uallicus)*	Australia.

BIRDS (cont'd.)

NAME	WHERE FOUND
Parrot, Imperial *(Amazona imperialis)*	West Indies: Dominica.
Parrot, Night *(Geopsittacus occidentalis)*	Australia.
Parrot, Orange-bellied *(Neophema chrysogaster)*	Australia.
Parrot, Red-browed *(Amazona rhodocorytha)*	Brazil.
Parrot, St. Lucia *(Amazona versicolor)*	West Indies: St. Lucia.
Parrot, St. Vincent *(Amazona guildingii)*	West Indies: St. Vincent.
Parrot, Scarlet-chested *(Neophema splendida)*	Australia.
Parrot, Thick-billed *(Rhynchopsitta pachyrhyncha)*	Mexico, United States.
Pelican, Brown *(Pelecanus occidentalis)*	Mexico, United States, Panama, Puerto Rico, etc.
Penguin, Galápagos *(Spheniscus mendiculus)*	Ecuador: Galápagos Islands.
Pheasant, Bar-tailed *(Syrmaticus humiae)*	Burma, China.
Pheasant, Blyth's tragopan *(Tragopan blythii)*	China, Burma, India.
Pheasant, Brown-eared *(Crossoptilon mantchuricum)*	China.
Pheasant, Cabot's tragopan *(Tragopan caboti)*	China.
Pheasant, Chinese monal *(Lophophorus lhuysii)*	China.
Pheasant, Edward's *(Lophura eduardsi)*	Vietnam.
Pheasant, Imperial *(Lophura imperialis)*	Vietnam.
Pheasant, Mikado *(Syrmaticus mikado)*	Taiwan.
Pheasant, Palawan peacock *(Polyplectron emphanum)*	Philippines.
Pheasant, Sclater's monal *(Lophophorus sclateri)*	China, Burma, India.
Pheasant, Swinhoe's *(Lophura suinhoii)*	Taiwan.
Pheasant, Western tragopan *(Tragopan melanocephalus)*	India, Pakistan.
Pheasant, White-eared *(Crossoptilon crossoptilon)*	China, Tibet, India.
Pigeon, Azores wood *(Columba palumbus azorica)*	Azores.
Pigeon, Chatham Island *(Hemiphaga novaeseelandiae chathamensis)*	New Zealand.
Piopio *(Turnagra capensis)*	Mascarene Islands: Réunion Island.
Plover, New Zealand shore *(Thinornis novae-seelandiae)*	New Zealand.
Rail, Auckland Island *(Rallus pegtoralis muelleri)*	New Zealand.
Rhea, Darwin's *(Pierocnenia pennata)*	Argentina, Peru, Uruguay, Bolivia.

BIRDS (cont'd.)

NAME	WHERE FOUND
Robin, Chatham Island *(Petroica traversi)*	New Zealand.
Robin, Scarlet-breasted *(Petroica multicolor multicolor)*	Australia: Norfolk Island.
Rock-Fowl, Gray-necked *(Picathartes oreas)*	Cameroon.
Rock-Fowl, White-necked *(Picathartes gymnocephalus)*	Togo to Sierra Leone.
Roller, Long-tailed ground *(Uratelornis chimaera)*	Madagascar.
Scrub-bird, Noisy *(Atrighornis clamosus)*	Australia.
Shama, Cebu black *(Copsychus niger cebuensis)*	Philippines.
Shrike, Mauritius cuckoo *(Coquus typicus)*	Mascarene Islands: Mauritius.
Shrike, Réunion cuckoo *(Coquus newtoni)*	Mascarene Islands: Réunion Island.
Starling, Ponape Mountain *(Aplonis pelzelni)*	Caroline Islands: Ponape Island.
Starling, Rothschild's *(Leucopsar rothschildi)*	Indonesia: Bali.
Stork, White oriental *(Ciconia ciconia boyciana)*	Japan, Korea, China, U.S.S.R.
Tern, California least *(Sterna albifrons browni)*	Mexico, United States.
Thrasher, White-breasted *(Ramphocinclus brachyurus)*	West Indies: Martinique, St. Lucia.
Trembler, Martinique brown *(Cinclocerthia ruficauda gutturalis)*	West Indies: Martinique.
Wanderer, Plain *(Pedionomus torquatus)*	Australia.
Warbler, Bachman's *(Vermivora bachmanii)*	Cuba, United States.
Warbler, Barbados yellow *(Dendroica petechia petechia)*	West Indies: Barbados.
Warbler, Kirtland's *(Dendroica kirtlandii)*	Bahamas, United States.
Warbler, Reed *(Acrocephalus luscinia)*	Mariana Islands.
Warbler, Rodrigues *(Bebrornis rodericanus)*	Indian Ocean: Rodrigues Island.
Warbler, Semper's *(Luecopeza semperi)*	West Indies: St. Lucia.
Warbler, Seychelles *(Bebrornis sechellensis)*	Seychelles.
Whipbird, Western *(Psophodes nigrogularis)*	Australia.
White-eye, Ponape great *(Rukia sanfordi)*	Caroline Islands: Ponape.
White-eye, Seychelles *(Zosterops modestus)*	Seychelles.
Woodpecker, Imperial *(Campephilus imperialis)*	Mexico.
Woodpecker, Ivory-billed *(Campephilus principalis)*	Cuba, United States.

BIRDS (cont'd.)

NAME	*WHERE FOUND*
Woodpecker, Tristram's *(Dryocopus javensis richardsi)*	Korea.
Wren, Guadeloupe house *(Troglodytes aedon guadeloupensis)*	West Indies: Guadeloupe
Wren, New Zealand bush *(Xenicus longipes)*	New Zealand.
Wren, St. Lucia *(Troglodytes aedon mesoleucus)*	West Indies: St. Lucia.

MAMMALS

NAME	*WHERE FOUND*
Anoa *(Anoa depressicornis)*	Indonesia.
Armadillo, Pink fairy *(Chlamyphorus truncatus)*	Argentina.
Ass, African wild *(Equus asinus)*	Ethiopia, Somalia, Sudan.
Ass, Asian wild *(Equus hemionus)*	Pakistan, Iran, India, China, Afghanistan, Central Asia.
Aye-aye *(Daubentonia madagascariensis)*	Madagascar.
Bandicoot, Barred *(Perameles bougainville)*	Australia.
Bandicoot, Desert *Perameles eremiana)*	Australia.
Bandicoot, Lesser rabbit *(Macrotis leucura)*	Australia.
Bandicoot, Pig-footed *(Chaeropus ecaudatus)*	Australia.
Bandicoot, Rabbit *(Macrotis lagotis)*	Australia.
Banteng *(Bibos banteng)*	Southeast Asia.
Bear, Mexican grizzly *(Ursus arctos nelsoni)*	Mexico.
Bison, Wood *(Bison bison athabascae)*	Canada.
Cat, Tiger *(Felis tigrina)*	Costa Rica to northern South America.
Cheetah *(Acinonyx jubatus)*	Africa, Asia Minor, India.
Colobus, Red *(Colobus bandius rufomitratus)*	Kenya.
Colobus, Zanzibar red *(Colobus badius kirkii)*	Tanzania: Zanzibar.
Deer, Bawean *(Helaphus kuhli,* or *Cervus kuhli)*	Indonesia.
Deer, Brow-Antlered, Eld's *(Cervus eldi)*	India, Southeast Asia.
Deer, Marsh *(Blastocerus dichotomus)*	Argentina, Uruguay, Brazil, Paraguay.
Deer, McNeill's *(Cervus elaphus macneilli)*	China, Tibet.
Deer, Persian fallow *(Dama dama mesopotamica)*	Iraq, Iran.
Deer, Swamp *(Cervus duvauceli)*	India, Nepal.

MAMMALS (cont'd.)

NAME	*WHERE FOUND*
Dibbler *(Antechinus apicalis)*	Australia.
Dog, Asiatic wild *(Cuon alpinus)*	U.S.S.R., India.
Dugong *(Dugong dugon)*	East Africa to Ryukyu Islands.
Ferret, Black-footed *(Mustela nigripes)*	United States, Canada.
Forester, Tasmanian *(Macropus giganteus tasmaniensis)*	Australia.
Fox, Northern kit *(Vulpes velox hebes)*	Canada.
Gazelle, Clark's (Dibatag) *(Ammordorcas clarkei)*	Somalia, Ethiopia.
Gazelle, Cuviers *(Gazella cuvieri)*	Morocco, Tunisia.
Gazelle, Mhorr *(Gazella dama mhorr)*	Morocco.
Gazelle, Moroccan Dorcas *(Gazella dorcas massaesyla)*	Morocco, Algeria.
Gazelle, Rio de Oro Dama *(Gazella dama lozanoi)*	Western Sahara.
Gazelle, Slender-horned (Rhim) *(Gazella leptoceros)*	Sudan, Algeria, Egypt, Libya.
Gibbon, Kloss *(Hylobates klossi)*	Indonesia.
Gibbon, Pileated *(Hylobates pileatus)*	Laos, Thailand, Cambodia.
Gorilla *(Gorilla gorilla)*	Central and Western Africa.
Hartebeest, Swayne's *(Alcelaphus buselaphus swaynei)*	Ethiopia.
Hog, Pygmy *(Sus salvanius)*	India, Nepal, Bhutan, Sikkim.
Hyaena, Barbary *(Hyaena hyaena barbara)*	Morocco.
Hyaena, Brown *(Hyaena brunnea)*	South Africa.
Ibex, Pyreanean *(Capra pyrenaica pyrenaica)*	Spain.
Ibex, Walia *(Capra walie)*	Ethiopia.
Impala, Black-faced *(Aepyceros melampus petersi)*	Southwest Africa, Angola.
Indris, Sifakas, Avahis *(Indriidae;* all members of the genera *Indri, Avahi, Indriidae;* all members of *Propithecus)*	Madagascar and Comoro Islands.
Jaguar *(Panthera onca)*	Central and South America.
Kangaroo, Eastern gray *(Macropus giganteus)*	Australia
Kangaroo, Red *(Megaleia rufa)*	Australia
Kangaroo, Western gray *(Macropus fulginosus)*	Australia
Kouprey *(Bos sauveli)*	Cambodia.

MAMMALS (cont'd.)

NAME	*WHERE FOUND*
Langur, Douc *(Pygathrix nemaeus)*	Indochina, China, and Hainan Island.
Langur, Pagi Island *(Simias concolor)*	Indonesia.
Lechwe, Black *(Kobus leche smithemani)*	Zambia.
Lemurs (Lemuridae; all members of the genera *Lemur, Hapalemur, Lepilemur, Cheirogaleus, Microcebus, Phaner*)	Madagascar and Comoro Islands.
Leopard *(Panthera pardus)*	Africa, Asia Minor, India, Southeast Asia, Korea.
Leopard, Formosan clouded *(Neofelis nebulosa brachyurus)*	Taiwan.
Leopard, Snow *(Panthera uncia)*	Central Asia.
Lion, Asiatic *(Panthera leo persica)*	India.
Lynx, Spanish *(Felis lynx pardina)*	Spain.
Macaque, Lion-tailed *(Macaca silenus)*	India.
Manatee, Amazonian *(Trichechus inunguis)*	Amazon Basin.
Manatee, West Indian (Florida) *(Trichechus manatus)*	Caribbean, northern South America.
Mangabey, Tana River *(Cercocebus galeritus galeritus)*	Kenya.
Margay *(Felis wiedii)*	Central and South America.
Marmoset, Goeldi's *(Callimico goeldii)*	Brazil, Colombia, Ecuador, Peru.
Marsupial, Eastern Jerboa *(Antechinomys laniger)*	Australia.
Marsupial-mouse, Large desert *(Sminthopsis psammophila)*	Australia.
Marsupial-mouse, Long-tailed *(Sminthopsis longicaudata)*	Australia.
Marten, Formosan yellow-throated *(Martes flavigula chrysospila)*	Taiwan.
Monkey, Red-backed squirrel *(Saimiri oerstedii)*	Costa Rica, Panama.
Monkey, Spider *(Ateles geoffroyi frontatus)*	Costa Rica, Nicaragua.
Monkey, Spider *(Ateles geoffroyi panamensis)*	Costa Rica, Panama.
Monkey, Woolly spider *(Brachyteles arachnoides)*	Brazil.
Mouse, Field's *(Pseudomys fieldi)*	Australia.
Mouse, Gould's *(Pseudomys gouldii)*	Australia.
Mouse, New Holland *(Pseudomys novaehollandiae)*	Australia.

MAMMALS (cont'd.)

NAME	*WHERE FOUND*
Mouse, Shark Bay *(Pseudomys praeconis)*	Australia.
Mouse, Shortridge's *(Pseudomys shortridgei)*	Australia.
Mouse, Smoky *(Pseudomys fumeus)*	Australia.
Mouse, Western *(Pseudomys occidentalis)*	Australia.
Native-cat, Eastern *(Dasyurus viverrinus)*	Australia.
Numbat *(Myrmecobius fasciatus)*	Australia.
Ocelot *(Felis pardalis)*	Central and South America.
Orangutan *(Pongo pygmaeus)*	Indonesia, Malaysia, Brunei.
Oryx, Arabian *(Oryx leucoryx)*	Arabian Peninsula.
Otter, Cameroon clawless *(Paraonyx microdon)*	Cameroon.
Otter, Giant *(Pteronura brasiliensis)*	South America.
Otter, La Plata *(Lutra platensis)*	Uruguay, Argentina, Bolivia, Brazil.
Planigale, Little *(Planigale subtilissima)*	Australia.
Planigale, Southern *(Planigale tenuirostris)*	Australia.
Porcupine, Thin-spined *(Chaetomys subspinosus)*	Brazil.
Possum, Mountain pigmy *(Burramys parvus)*	Australia.
Possum, Scaly-tailed *(Wyulda squamicaudata)*	Australia.
Prairie Dog, Mexican *(Cynomys mexicanus)*	Mexico.
Pronghorn, Peninsular *(Antilocarpa americana peninsularis)*	Mexico.
Pronghorn, Sonoran *(Antilocapra americana sonoriensis)*	Mexico, United States.
Quokka *(Setonix brachyurus)*	Australia.
Rabbit, Volcano *(Romerolagus diazi)*	Mexico.
Rat, False water *(Xeromys myoides)*	Australia.
Rat, Stick-nest *(Leporillus conditor)*	Australia.
Rat-kangaroo, Brushtailed *(Bettongia penicillata)*	Australia.
Rat-kangaroo, Gaimard's *(Bettongia gaimardi)*	Australia.
Rat-kangaroo, Lesueur's *(Bettongia lesueur)*	Australia.
Rat-kangaroo, Plain *(Caloprymnus capestris)*	Australia.
Rat-kangaroo, Queensland *(Bettongia tropica)*	Australia.

MAMMALS (cont'd.)

NAME	*WHERE FOUND*
Rhinoceros, Great Indian *(Rhinoceros unifornis)*	India, Nepal.
Rhinoceros, Javan *(Rhinoceros sondaicus)*	Indonesia, Burma, Thailand.
Rhinoceros, Northern white *(Ceratotherium simum cottoni)*	Zaire, Uganda, Sudan, Central African Republic.
Rhinoceros, Sumatran *(Didermoceros sumatrensis)*	Southeast Asia—Bangladesh to Vietnam to Indonesia (Borneo).
Saki, White-nosed *(Chiropotes albinasus)*	Brazil.
Seal, Mediterranean monk *(Monachus monachus)*	Mediterranean, Northwest African Coast and Black Sea.
Seladang (Gaur) *(Bos gaurus)*	India, Southeast Asia, Bangladesh.
Serval, Barbary *(Felis serval constantina)*	Algeria.
Shou *(Cervus elaphus wallichi)*	Tibet, Bhutan.
Sloth, Brazilian three-toed *(Bradypus torquatus)*	Brazil.
Solenodon, Cuban *(Atopogale cubana)*	Cuba.
Solenodon, Haitian *(Solenodon paradoxus)*	Dominican Republic, Haiti.
Stag, Barbary *(Cervus elaphus barbarus)*	Tunisia, Algeria.
Stag, Kashmir *(Cervus elaphus hanglu)*	Kashmir.
Tamaraw *(Anoa mindorensis)*	Philippines.
Tamarin, Golden-rumped (Golden-headed tamarin; Golden lion marmoset) *(Leontideus,* all species)	Brazil.
Tapir, Brazilian *(Tapirus terrestris)*	Venezuela, Argentina, Brazil, Columbia.
Tapir, Central American *(Tapirus bairdii)*	Southern Mexico to Colombia and Ecuador.
Tapir, Mountain *(Tapirus pinchaque)*	Colombia, Ecuador, Peru.
Tiger *(Panthera tigris)*	Central Asia, China, and Korea to India, Indonesia, and Malaysia.

MAMMALS (cont'd.)

NAME	*WHERE FOUND*
Tiger, Tasmanian (Thylacine) *(Thylacinus cynocephalus)*	Australia.
Uakari *(Cacajao,* all species)	Peru, Colombia, Brazil, Venezuela, Ecuador.
Vicuña *(Vicugna vicugna)*	Peru, Bolivia, Argentina.
Wallaby, Banded hare *(Lagostrophus fasciatus)*	Australia.
Wallaby, Brindled nail-tail *(Onychogalea frenata)*	Australia.
Wallaby, Crescent nail-tail *(Onychogalea lunata)*	Australia.
Wallaby, Parma *(Macropus parma)*	Australia.
Wallaby, Western hare *(Lagorchestes hirsutus)*	Australia.
Wallaby, Yellow-footed rock *(Petrogale xanthopus)*	Australia.
Whale, Blue *(Balaenoptera musculus)*	Oceanic.
Whale, Bowhead *(Balaena mysticetus)*	Oceanic.
Whale, Finback *(Balaenoptera physalus)*	Oceanic.
Whale, Gray *(Eschrichtius gibbosus)*	Oceanic.
Whale, Humpback *(Megaptera novaeangliae)*	Oceanic.
Whale, Right *(Eubalaena)* spp	Oceanic.
Whale, Sei *(Balaenoptera borealis)*	Oceanic.
Whale, Sperm *(Physeter catodon)*	Oceanic.
Wolf, Maned *(Chrysocyon brachyurus)*	Brazil, Bolivia, Paraguay, Argentina.
Wombat, Barnard's *(Lasiorhinus barnardi)*	Australia.
Wombat, Queensland hairy-nosed *(Lasiorhinus gillespiei)*	Australia.
Yak, Wild *(Bos grunniens mutus)*	Tibet, India.

Index